Adam's Creatures, or *The Book of Robots*:
The Rise of Robots, Intelligent Agents and Machines that Learn from Humans. Second Edition

ii

Robert J. Betts
http://www.RobertBettsHomePage.weebly.com

Adam's Creatures, or *The Book of Robots*:
The Rise of Robots, Intelligent Agents and Machines that Learn from Humans. Second Edition

Adam's Creatures, or *The Book of Robots*:
The Rise of Robots, Intelligent Agents and
Machines that Learn from Humans. Second Edition.
©2017, by Robert J. Betts.
ISBN-13: 978-1542942041
ISBN-10: 1542942047
Printed by CreateSpace (www.createspace.com), an Amazon.com Company.
Betts, Robert J.

Dedication

For my mother (1923–2012), who taught me endurance under suffering, and for my father (1915–1974), who taught me courage under fire.

Remember, that I am thy creature; I ought to be called thy Adam.
From *Frankenstein, or the Modern Prometheus,*
by Mary Wollstonecraft Shelley.

About the Author

Robert J. Betts (BA, MSc) is an operations research analyst, programmer in numerical computing (C, Fortran, MATLAB) and a software systems engineer at a small IT startup. Mister Betts is a published author whose interests in applied computer science include machine learning, discrete mathematics applications and cryptographic algorithms. For the last fourteen years he also has been a volunteer tutor in undergraduate mathematics.

In 2002 Robert Betts graduated with a Bachelor of Arts degree in mathematics from the University of Massachusetts Boston and his minor was computer science. He has a certificate in the Unix operating system (2012) from the Department of Mathematics and the Division of Continuing Studies at the University of Massachusetts Lowell. He has more than thirty-seven years of experience in writing Unix/Linux shell scripts and in writing programs in FORTRAN 77 on Unix operating systems and mainframes, and writing code in C, Fortran 90, Perl and Python on Unix System V, Ubuntu Linux and on Windows for various projects in computational mathematics, science and engineering.

In 2015 Mister Betts completed his graduate studies in information technology at Southern New Hampshire University, where he completed a software engineering related capstone project to fulfill requirements for the Master of Science degree with a specialization in relational database theory, design and database development.

Supplementary graduate school course material that he also has covered in applied mathematics and computer science at UMASS Lowell, SNHU and at other universities include: Data mining, Java programming, computer and network security, computer simulation and animation with Flexsim, operating systems, data communications, linear and nonlinear dynamical systems, bifurcation and chaos, random processes, experimental design and probability and statistics.

Robert J. Betts is a member of The Association for Computing Machinery and The Mathematical Association of America. He enjoys reading literary classics, chess, watching old television shows like *Gunsmoke*, *The Rockford Files* and *Star Trek: The Next Generation* and listening to Bach, Beethoven, Brahms, Bruckner, Max Reger and Gustav Mahler.

x

Contents

Preface to the Second Edition

This Second Edition contains some additional material in comparison to the free pdf edition at Scribd (www.scribd.com).

This is not a textbook on robotics or on robotics simulation. Rather it is a book about robots, that is, robots and automatons in Western mythology, in history, in science fiction and fantasy and particularly within the technological realities of our modern civilization.

This book is an exploratory journey through the clever and inventive mind of *homo sapiens*, when it comes to the human development of myths, ideas, literature and physical inventions that relate to the creation of the most incredible and imaginative, mechanical facsimiles of human men and women down through the ages that have come to be called robots.

From the bronze automaton Talos in the *Argonautica* to the formidably powerful, alien robot Gort in *The Day the Earth Stood Still*, mechanical men, women and various robot machines have fascinated one human visionary after another across a diverse range of fields and genres: The mechanical doll of Descartes, the violent monster created by Mary Shelly's Viktor von Frankenstein, the fearsome humanized robots in *Blade Runner*, the inquisitive android Data aboard the starship *Enterprise*, the Mars Pathfinder, Grace and Ada, Asimo and Atlas. This book, as an excursion into ideas about robots old and new, presents an amalgamation of various inventive ideas and facts about robots that have appeared respectively, within gothic fiction, science fiction and fantasy, within astronomy and as well as in cybernetics, bioengineering, information technology[1] and within artificial intelligence. All these multifarious human worlds of ideas and inventiveness about robots and technology in myth, in folklore, in fiction and even in hi-tech reality, in a sense, are interdependent.

In modern times robots have been the abundant stuff of science fiction and fantasy for the last six or seven decades, particularly after technological advances led to Sputnik, Telstar and the transistor. Robot space probes from Sputnik to New Horizons have done everything from circumnavigating the globe in low Earth orbit to transmitting precious data about Pluto's surface features back to the Jet Propulsion laboratory at NASA.

Perhaps at least some chapters in this book are best suited for readers who might have some technical professional background or at least have interests in technical things. That classification deliberately is very broad and general to include everyone from robotics hobbyists, IT bloggers, college students in STEM programs, science fiction writers and enthusiasts of the genre, software engineers and computer programmers, database developers with an interest in decision support systems and knowledge discovery in databases and IT people who have professional interests in data mining, pattern analysis, machine

---

[1]Today information technology, which includes data communications, telecommunications, computer hardware and the software associated with it, arose as a sort of offspring of electrical engineer Claude Shannon's development of information theory and the theory of "formal languages" one studies in computer science, which we shall see in a later chapter.

learning and data science. Frankly *anyone* who thinks robots and artificial intelligence are intriguing but who also are interested in the serious moral and ethical challenges that humanity will confront due to the future uses of intelligent robots and AI, can find something in this book that will nurture in them further interest in the subjects of robotics and artificial intelligence. It also might interest those who have valid concerns about the possible dangers that humanity will confront if robots and artificial intelligence are used by human beings, corporations or nations in the wrong ways. Any readers with some familiarity with sets, precalculus, some elementary probability and calculus ought to be able to navigate through those chapters with some mathematical content, such as chapters 4, 7 and 10. Those who have had an introductory course in computer programming or scripting should be able to follow the pseudocode in Chapter 6 and Chapter 12 with little difficulty.

Since this book is meant to accommodate those readers who might not have an educational or professional background in science, technology and mathematics disciplines we trust that professional roboticists, mechanical engineers, electrical engineers, computer scientists and their students will be reminded of this fact when one does not see more explicit material covered on robotics and AI, such as more explicit material on specific supervised and unsupervised machine learning algorithms, knowledge bases and ontologies, minimax, alpha-beta pruning and heuristic search trees such as Arthur Samuel used in his famous Checkers Playing program, or more technical details about servomechanisms, actuators, particle filters, probabilistic robotics, etc., in the chapters. Most of these more involved topics are beyond the scope of this book. A Bibliography is provided so that any interested and courageous readers who are inclined more to pursue technical topics or mathematics can find material that includes more detailed material about AI and robots.

The book has two parts to it. Most of the topics in Part I: "Robots in Mind," have to do with the imaginative conceptions about mechanical men and women as conceived by myth, folklore and by various writers and film directors past and present. It includes also very early attempts to construct artificial human beings and other automatons. Part II: "Robots in Matter" presents actual developments in robotics and artificial intelligence as this has happened within the areas of mechanical and electrical engineering and computer science in particular since the end of World War Two.

Part II bridges several science and technology disciplines: Information theory, computer science both theoretical and applied, information technology, cybernetics, machine learning, artificial intelligence, game theory, computational geometry and programming.

Robotics and AI are two fields that combine various scientific and technological disciplines that are interconnected and not disciplines that remain within mutually exclusive silos. The programmers might code for robotic simulations but they need the mechanical engineers before their robot software simulation can be tested on real robot actuators. Robots cannot learn without first having access to various machine learning and artificial intelligence algorithms that are derived by computer scientists and persons with a background in statistics or in statistics applications. The programs that run those algorithms behave like finite state machines, a topic that derives from theoretical computer science. Swarm robots in the laboratory or in the field on land can use a wireless network for their secure communications. Computer networks and their various protocols of course are familiar things to those with a background in information technology.

One ought to note also that even an automated space probe or satellite launched into deep space cannot communicate data back to Earth across a communication channel without first having some way to correct errors transmitted in the data stream, something that was studied extensively by Claude E. Shannon, one of the early pioneers in the mathematical theory of information.

Although this book will not instruct the reader on how to build a robot arm or virtual human avatar that uses the newest natural language processing algorithms like Hidden Markov Models or deep learning, one will read how robots and mechanical automatons of different kinds have fascinated humans from René Descartes's automaton Francine to the android Data in the Star Trek mythology conceived by Gene Roddenberry. There has been a span of time more than four centuries long between the simple automatons of René Descartes and the expansive ideas of Alan Turing, John McCarthy and Marvin Minsky. Nevertheless this book will strive to cover things without going into every minute detail.

Those "Baby Boomers" who recall vividly the tense years of the Cold War (as does this author, in fact) will recall as well the awe they felt when they went to the cinema to watch Gort disintegrate an army tank on screen, or to see Tobor the Great defeat Communist spies and nefarious saboteurs. Even from the nineteen sixties until today robots have kept us enthralled as we watched them on television, on DVDs, video games and in movie theaters. TV and film characters like Robby the Robot, Star Trek's Data and the Terminator still elicit interest among those laypersons young or old who are driven by fascination to wonder what robots can do now and what they might be able to do in the future. One will read this book one hopes, to get a sense for all the collective human fuss and wonderment about robots, about what they are capable of learning and doing, but also why some men and women today in the fields of pure science and philosophy are concerned very deeply about what it might portend when robots and machines will surpass even humans in intelligence.

It is excruciating to prepare a book with some GUI based word processing programs along with any required graphics images, formatting, etc., then to use more GUI applications to convert a document in a word processing format along with .jpeg files into .epub or .mobi files for upload to some conventional epublishers, at least in the opinion of this author. Also it can be agonizing trying to format mathematical equations in a word processing document along with graphics, etc., then to try to reformat all that with the GUI applications used by most online epublishers into .epub and .mobi files. Moreover many .pdf to .epub file conversion programs frequently do not render page numbers, images, scientific diagrams and mathematical equations correctly because they place very restrictive demands on formatting. I considered converting the entire TeX/LaTeX file into HTML at the command line with an open source application, then using another open source application to convert the output HTML file into EPUB; but this then presented the problem of converting math equations and formulas properly along with in-line math expressions, from TeX/LaTeX into MathML. All this just proved to be far too tedious at this time. Was it all just too difficult for you? One might ask. No. There is a world of difference between technical difficulty and sheer tedium.

Therefore as an alternative this edition was written, prepared and typeset entirely with TeX/LaTeX using MikTeX 2.9, then output into PDF format. The final book was prepared as hardcopy by CreateSpace.

One Wiki book in particular was exceedingly helpful, that is, along with the well known textbook *LaTeX: A Document Preparation System User's Guide and Reference Manual*, by Leslie Lamport:

http://en.wikibooks.org/wiki/LaTeX/

Years ago the electrical engineer and AI researcher Arthur Samuel also made extensive contributions in the development of TeX documentation. His other noteworthy contributions in getting computers to beat humans at competitive games like checkers are discussed in Chapter 11 of this book.

There seems to be an unyielding determination among major online retailers not to sell ebooks online as pdf files. The original edition of this book was published as a pdf file on Scribd and it contains some color images. Most of the Wiki images for the Scribd edition were based on JPEG files subject to lossy data compression. Any increase in the size of these images for the hardcopy version caused data loss through the introduction of statistical noise which kept adding blur to the final images. Regrettably increasing the DPI in all the images by a professional graphics artist just was not an option. Book printing (i. e., as hardcopy) considerations along with various printing restrictions, specifications, etc., combined with the fact that this book is part of a nonprofit fundraising effort on the part of the author, would have made this entire publishing project infeasible if color images were included in this Second Edition, as it would have forced this author to offer the book to the public only at a ridiculously high price. Therefore this author regrets very much that all Wikimedia color images that were meant to be included in this hardcopy book are referenced only, since printing restrictions and certain file specifications made it impossible to include any color images at this time. So the actual Figures for the Wiki images that are mentioned or referenced in this book can be found at Wikipedia.

The following will serve as a kind of de facto Chapter to Chapter Figure Key:

"Talos," Chapter 1: http://en.wikipedia.org/wiki/File:Didrachm_Phaistos_obverse_CdM.jpg
"The Golem," Chapter 1: http://en.wikipedia.org/wiki/File:Golem_1920_Poster.jpg
"The Scream," Chapter 2: http://en.wikipedia.org/wiki/File:The_Scream.jpg
"Tik-Tok," Chapter 2: http://en.wikipedia.org/wiki/File:Tiktok.png
"R. U. R.," Chapter 2: http://en.wikipedia.org/wiki/File:Wpa-marionette-theater-presents-rur.jpg
"Ada Lovelace," Chapter 3: http://en.wikipedia.org/wiki/File:Ada_Lovelace.jpg
"Explorer," Chapter 5: http://en.wikipedia.org/wiki/File:Explorer1.jpg
"Viking," Chapter 5: http://en.wikipedia.org/wiki/File:Viking_Orbiter_releasing_the_lander.jpg
"Pathfinder," Chapter 5: http://en.wikipedia.org/wiki/File:Pathfinder01.jpg
"Elektro and Sparko," Chapter 7: http://en.wikipedia.org/wiki/
File:Senator_John_Heinz_History_Center_-_IMG_7802.JPG
"BigDog," Chapter 9: http://commons.wikimedia.org/wiki/
File:Bio-inspired_Big_Dog_quadruped_robot_is
_being_developed_as_a_mule_that_can_traverse_difficult_terrain.tiff
"Carl Sagan," Chapter 15: http://commons.wikimedia.org/wiki/
File:Carl_Sagan_Planetary_Society.JPG

All Wiki images that are referenced in this publication are for nonprofit and for educational purposes.

One dollar and ninety-nine cents (1.99 US) from each and every purchase of this book either from any Amazon store or from eStore will be donated to the Wikimedia Foundation, Inc., which I found personally to be an extremely valuable repository of educational information when I was a graduate student in computer science and then in information technology. Whether this donation amount is increased or not in the future depends upon the number of book purchases. For any who might not know otherwise, Wikipedia articles do get revised frequently for improved accuracy when necessary by Wiki editors.

An additional one dollar and ninety-nine cents from each and every purchase of this book either from any Amazon store or from eStore will be donated to the World Wildlife Fund (www.worldwildlife.org) toward their efforts to end the senseless poaching of elephants and rhinos.

All other funds from any sales of this book, that is any other royalties that are left after the two aforementioned donations, are earmarked for a nonprofit startup which has a focus on providing free and low cost training and educational resources (MOOC) in the areas of Linux (LAN/WLAN/LDAP/Apache Server) system administration and computer and network security, and to provide funds for research and development in various areas of applied computer science.

Neither the World Wildlife Fund, Inc., nor The Wikimedia Foundation, nor Boston Dynamics, Inc. nor any of their employees, has endorsed this publication or any viewpoints of the author that might be contained within it. Nor has this book been published for the purpose of any personal remuneration either from these organizations or from any other corporation mentioned within this publication and that either is or is not engaged in research and development in robotics or in artificial intelligence. Nor has this book been written for such a purpose or intent. Nor has anyone from these aforementioned nonprofits or corporations requested that I raise donations for them. Furthermore no university, company or corporation mentioned in this publication that is promoting research and development either in robotics or in artificial intelligence has endorsed this publication or offered any remuneration for its publication.

Although the author of this book has volunteered on occasion as a Wiki editor, the author of this book neither is a paid employee of the Wikimedia Foundation nor is he a board member of its nonprofit corporation.

Back in the nineteen eighties and nineties the author had been a member of The Planetary Society. However neither The Planetary Society nor anyone employed with this organization has endorsed this publication nor has anyone from this nonprofit offered or paid any remuneration to the author.

The Figures included in this book were created with an open source CASE (*C*omputer *A*ided *S*oftware *E*ngineering) software application. All Figures had to be rendered in black and white.

Seldom does a new book get published without typos creeping somewhere within the pages. In fact the number of typos on any given page of a newly typeset book can be Poisson distributed given the right value for the real parameter λ. If this turns out to be the rule rather than the exception with this publication the author extends his humble apologies. Also one might note there is no List of Figures. It is the intention of this author to resolve this lack of a List of Figures along with any typos and/or graphics issues in a second revised edition.

If intelligent robots and machines truly are to be a part of humanity's future it behooves humanity to prepare for that by advancing not only new technologies and in the knowledge of how to use them, but also within the areas of wisdom, enlightenment, ethics, sound judgement and moral decency.

Robert J. Betts
January, 2017

Introduction

Like father, like son. Like mother, like daughter.

Frequently very young children love to imitate their parents. In fact many parents even have wished that their son or daughter would grow up to be just like "daddy" or "mommy." For instance one can wonder whether or not the naturalist Charles Darwin had wished that his son grow up to become a naturalist as he had been. At the very least his son did become a physicist.

Even in the Genesis account we read *And God created man in his own image, in the image of God created he him; male and female he created he them*[2]. In the account God even did not want his new creatures, his children, to be created in the image of something that did not remind him of himself.

In our highly complex, day to day world though, as a parent one sees that things do not work out for children always as one would like them to work out. Sometimes the father is an abusive bully and the son grows up to become like him. A mother might abuse drugs and to her misfortune she might see that her adult daughter years later, is doing likewise. Should we think that intelligent robots always would behave differently in the future if they are exposed to the darker side of human behavior? Not necessarily. Frequently intelligent offspring do learn things by imitating their parents. That is true for bear cubs and it also can be true for human children and for highly intelligent robots in the future.

Yet robots are fascinating, especially ones that can walk or talk. They too are the offspring of their parents, roboticists and computer scientists who watch them come to birth not inside a maternity ward but inside a laboratory. Parents of human children wish their children to grow up like them. Roboticists wish their creations to function well after development with minimal error and with high fault tolerance. But robots even can begin to imitate humans if their creators, their parents, develop them to do so.

From the past to the present many different kinds of robots have been conceived as well as created from Maria[3] to iCub. Lang's robot *Doppelgängerin* character Maria was capable of imitating humans to the point of deceiving them successfully to turn against each other in violence. Will our real intelligent robots in the future be able to do something similar or worse? In Part II of this book we shall delve into this important issue.

Part I: "Robots in Mind," is a focus on how the various ideas about mechanical men and women have evolved through the course of human history. Chapter 2 for instance will present us briefly with some early automatons that were developed by René Descartes and others. Although these automatons lacked the sort of sophistication that Asimo or BigDog might have today, they are good examples of what inventive humans can do even with more primitive technology such as gears and cogwheels. Chapter 3 explores the hor-

---

[2]From Genesis Chapter 1, the American Standard Version
[3]From the German film *Metropolis*, by Fritz Lang

ror and science fiction ideas on robots in legend such as *The Golem* as well as by writers, playwrights and directors and producers of some popular films and television shows. Isaac Asimov for example was one of the first science fiction authors to explore the concept of humans having sex with robots. Ray Bradbury gave us a robot character conceived in the image and likeness of a caring, loving and human elderly grandmother. Other writers and directors treated more different themes, such as Karel Čapek, who with his *R. U. R.*, considered the possibility of the overthrow of human rule by biological facsimilies of humans. On the other hand ideas on the rights of robot replicants are treated in movies like *Blade Runner*. German film director Fritz Lang presented us with a scenario that illuminates us on how robots can be evil and can be used to exploit human evil and human weakness. Gene Rodenberry presents us with an ultimate computer so capable of thinking like a human that it kills humans. Arthur Clarke's HAL 9000 computer did something similar.

Chapter 4 and Chapter 5 give us robots of a different mind. we discuss there the work of Charles Babbage and Ada Lovelace on the "Babbage Engine" and the burgeoning new ideas of Alan Turing, Alonzo Church and others about machine computation, formal languages and machines that can think.

We begin Part II: "Robots in Matter," with a discussion on NASA's various robotic space probes from the early Pioneer missions to New Horizons and that have been developed after the end of World War II. Many of these space probes were highly successful autonomous machines and they reveal to us how beneficial robots can be when they are put to good use, even when they do not walk or talk. Chapters 7 and 8 deal with cybernetics, analogue robots and some of the pioneers who made contributions in these areas like William Grey Walter and Claude E. Shannon. Also we discuss the impact that information theory as developed by Shannon, has had on technological development.

Chapters 9 and 10 introduce us to other kinds of robots that have been developed other then space probes and after the development of the transistor and digital technology, robots that for example can roll, fly and walk which includes autonomous drones, Atlas and Asimo. Chapter 11 exposes us to what has been done already in bionics, using ideas from robotics to enable people with missing limbs to walk and to reach for a glass of water in ways that outdate the older more primitive prosthetics devices from a previous generation. Then Chapter 12 covers artificial intelligence as this has evolved since the early work of the two intellectual giants in the field, namely John McCarthy and Marvin Minsky.

Since this is not an instructional book on robotics one should not expect to find a lack of support for some overall thesis about humans, new technologies across the span of time and how humans interacting with various new technologies has impacted history. Even Mary Shelley's gothic horror novel *Frankenstein* offers some warning on how the human urge to create life can bring monsters into the world. But can robots become monsters if they learn the wrong things from men and women? All one has to do is to take a good and impartial hard look at human history to answer this question.

The reader should be aware that this book does not restrict itself to a discussion solely about robots and artificial intelligence in fiction and in fact. It also considers the development of new technologies over time within an historical context and how humans have adapted and used various new technologies for good or for ill.

As new technologies develop over time, so does the human capacity for causing death, chaos and destruction. In the battle of Kursk, USSR during World War Two the tank then caused more death and destruction than a bronze spear ever did in the hands of an ancient Babylonian soldier somewhere in Mesopotamia. The last three chapters of Part II, Chapters 13–15, serve as a caveat so to speak, since they discuss the potential dangers of AI and robotics that are possible due to unrestrained and unchecked, egregious human behaviors based on emotions such as egotism, anger, fear, hatred and greed.

What sort of behaviors?

Across America we Americans are divided violently by politics, by race, religion and nationality, even at the very time when the nation faces still a serious ongoing threat from ISIS, a terrorist organization that promotes and supports here and abroad the type of hatred and bigotry that ironically, millions of people in our country promote and support actively behind the shields of "political incorrectness" and "free speech." After all is said and done, how much difference is there anyway between an ISIS terrorist bigot somewhere in Syria or Iraq shouting that America is the "Great Satan," and an American bigot in New York screaming a racial slur and other offensive remarks about blacks, Hispanics or Muslims, before he or she commits a hate crime? Which of these two bigots is the true moral superior of the other? After September 11, 2001, President George W. Bush then reminded the nation that our war is with Al Qaeda, not with Islam. In his second inauguration address on January 20, 2005, President Bush reminded us of some words by Abraham Lincoln, someone who was very familiar with the offensive activities and demeaning words of the Anti-Irish, Anti-Catholic Know-Nothing bigots of his own era: "Those who deny freedom to others deserve it not for themselves."

Human prejudice frequently does have a direct impact on how technology is used. In the exhaustive tome *The Nature of Prejudice*, Harvard psychologist Gordon W. Allport described how bigotry and hate begin with name calling, progress onward to legal discrimination and proscription and end up with things like genocide and lynching. In the nineteen twenties and then the thirties the Nazis in Germany used the technology of printing presses and radio to spread lies and misinformation. They used new technology to invade countries with the use of new bombers and tanks and to engage in mass murder in the death camps. So one can imagine what the Wehrmacht, the SS and Doctor Göbbels would have done with things like computers, robots, artificial intelligence, social networking, the Internet and nuclear fission, if Adolph Hitler had listened more to physicist Werner Heisenberg and had learned more about nuclear fission, instead of basing his perspectives about science on racial theories. Nazi Germany's V2 rocket in fact was not too dissimiliar to the autonomous modern ICBM. But the same kinds of technologies that the Nazis had deployed to devastate places like London, Warsaw and Leningrad during the Second World War also was used to devastate Berlin and Dresden.

Daily in this country alone we see children and young adults driven to suicide or to near suicide through anonymous cyber bullying in Internet chatrooms and social networks. With regard to cyber bullying, the new First Lady Melania Trump, previously and before the election last year back in November, 2016 during the campaign when in Pennsylvania had remarked, "Our culture has gotten too mean and too rough....We have to find a better way to talk to each other, to disagree with each other, to respect each other....We need to teach our youth American values–kindness, honesty, respect, compassion, charity, understanding, cooperation....We must come together as Americans."

This is good advice for the entire nation. More than likely it also is the sort of advice the Reverends Billy Graham and Martin Luthor King if they both were alive today would have admonished *all* American citizens to do. But if two hundred and forty-one years of a shared American history, Judeo-Christian ethics, constitutional rule of law, the American systems of education, justice and jurisprudence and the Sermon on the Mount along with the sermons of Americans like Martin Luthor King and Billy Graham have not succeeded in getting all Americans to be kind to one another, respectful, compassionate, etc., then it is not likely the First Lady's timely words will be able to do it. Almost two thousand years ago Paul in one of his epistles admonished all Christians, "Let no corrupt speech proceed out of your mouth. Let all bitterness, and wrath, and anger, and clamor, and railing, be put away from you, along with all malice, and be ye kind to one another." However today this hardly is the sort of behavior practiced by millions of nominal Christians from coast to coast across America. That is true from the "mighty mountains of New York" to "the Stone Mountain of Georgia," *regardless of one's race, political party, sex or social status.* So a white bigot who uses racial slurs and hate speech in New York or California is no more a follower of Christ than is a black criminal who helps to burn and to loot a shopping mall in Missouri or Maryland. *For whoever shall keep the whole Law, but shall fail in one point, has become guilty of all.*

Recall too that centuries ago people like Aaron Burr, Andrew Jackson, Kentucky Congressman Henry Clay and his lawyer opponent Humphrey Marshall settled their political and personal disputes through more violent means.

Like it or not it is just a basic part of human nature to be self-centered, intolerant, judgemental and cruel, which is not true yet for robots. In his play *The Tragical History of Doctor Faustus* (a free copy is available online at Project Gutenberg) Christopher Marlowe has his protagonist Faust quote then renounce the following Johannine passage from the Vulgate Bible by Jerome:

> Si pecasse negamus, fallimur, et nulla est in nobis veritas.

Well, based upon the behaviors of billions of humans today on "Spaceship Earth," Faust must have been right to renounce. Are not all the sinners, wrongdoers and troublemakers *always* inside the other political party, race, nation, religion, sex or tribe? The result is that sheer might and power give one the right to cast the first stone. After all, when have the members of a lynch mob or criminal street gang ever considered themselves to be guilty of anything? In their own eyes they too have done nothing wrong. So they go home with clean hands after their Sunday evening church barbecue or after their latest successful convenience store robbery.

All human societies, especially those societies that are advanced and highly complex, are a reflection of the very best and the very worst aspects of human nature. Therefore one cannot transform human societies for the betterment of all in them unless one has the power also to eliminate first the very worst of human traits: Egotism and hubris, self-righteousness, groupthink, bigotry of various kinds, hatred, violence, aggression and irrationality being among some of these worst more evil human traits. It is sad but true that no religion has been able to do this in the last six thousand to ten thousand years of human history. One could get the impression that humanity has allowed itself to be duped and swindled by one religion after another for at least six millennia while some kind of cosmic but undetected Mephistopheles enjoys his good laugh at the stupid and deluded human race.

Only Jesus was a genuine Christian. The words of Nietzsche are apropos: *Im Grunde gab es nur einen Christen, und der starb am Kreuz.* Every other Christian either is just an imitator of Jesus the Nazarene or else a hypocrite who enjoys the trappings of some organized religion. In fact after one has read the words and writings of Jesus in the synoptic texts, of Paul, Peter, James, John, Irenaeus and Ignatius, it would be impossible for any rational and objective person to accept the premise that hundreds of millions of people anywhere in the world who tolerate cyber bullying, racial and religious hatred and hate speech, race riots, political corruption, police brutality, violent protest and willful vandalism on college campuses, violent crime within their own communities whether those communities are rich or poor and homelessness, are a people who comprise a community or nation of *Christians*, in which there is "liberty and justice for all."

One of the most tragic things about life in America today as well as elsewhere is that many have abandoned concepts like wisdom, morality, human decency, fair play, impartial justice and ethics for prejudicial 'yardsticks' or litmus tests they use instead, to determine whether some course of moral or ethical action is "liberal" or "conservative." Yet it is hard to imagine that Americans like Abraham Lincoln, Henry David Thoreau, US Grant, Calvin Coolidge, Herbert Hoover, John Muir, Eleanor Roosevelt, Dwight D. Eisenhower, Robert Taft, Thomas Dewey, Prescott Bush, Margaret Chase Smith, Barry Goldwater, Lyndon Johnson, Lady Bird Johnson, Rachael Carson, John F. Kennedy, Edward Kennedy, Jackie Robinson, Viola Liuzzo, Martin Luthor King Jr., Gerald Ford, Betty Ford, Nelson Rockefeller, Admiral Hyman Rickover, Jack Kemp, Thomas "Tip" O'Neill, Roberto Clemente, the American architect Fazlur Khan, musician Selena Quintanilla, neurosurgeon Ayub Ommaya, Senator Daniel Inouye, President Ronald Reagan or even Jesus the Nazarene and Paul the Apostle for that matter, would have approved of such an absurd and limited human perspective. As someone who was a kid in the nineteen fifties and a growing teenager in the sixties, I recall very well what Americans like Dwight Eisenhower, Herbert Hoover, Margaret Chase Smith[4], Gerald Ford, Betty Ford, Barry Goldwater and Nelson Rockefeller cherished as American ideals and what bad human traits they detested to see and to hear whenever such bad traits were manifested in public, since back then they were alive and they spoke for themselves frequently in newspapers, in printed books, on radio and on television.

All these Americans knew that Judeo-Christian values, fair play, impartial justice and the "equal protection of the laws" are not things that ought to depend upon the race, religion, economic status or political party to which a citizen belongs, except perhaps within a totalitarian state or under lynch law rule by a mob. They also knew that the right to free speech did not mean every American has to accept insulting or demeaning language without outrage, rejection or protest. Back in the nineteen thirties the German American Bund learned that fact the hard way in Februrary, 1939 when during one of their rallies, they faced angry protesters outside New York's Madison Square Garden. Prior to that the anti-black, anti-immigrant Ku Klux Klan had learned it in the nineteen twenties when they found out they had no listening ear or sympathy in the White House of Calvin Coolidge.

So perhaps, since one cannot expect that every human being who has access to Internet technologies will "behave," the best way to fight cyber bullying and to block the posts of cyber bullies in particular in the short term at least, is with new software user applications, based on artificial intelligence or machine learning, that can block automatically any malicious posts even as they are posted but that are independent from any web server applications that are run by the social media website. It was a similar approach to software development that led to combatting email spam.

Today many Americans are judged on sight not by the content of their character but by the color of their skin, by the practice of their religion and by the practice of their language at birth. Internet technologies are being used today to spread lies, false sto-

---

[4]Miss Smith was a widowed Republican congresswoman when I was a kid. She had spoken out courageously against the vicious tactics of Senator Joseph McCarthy as did others.

ries, hatred and slander, just as Paul Joseph Göbbels used the printing press and radio to spread lies to the German people. Human prejudice and name calling is one thing. Frequently though name calling seldom is enough against the scapegoat. Unrestrained anger, violence and hatred against an out-group can make civil rights and constitutional laws meaningless in the United States of America just as such laws had become meaningless in the Democratic Republic of the Congo during the Congolese tribal wars after 1996. Such things threaten to unravel the very fabric of American civilization if it is left unchecked for too long a time, until societal dysfunction becomes fatal and irreversible. So the real problem is not technology, but bad human behaviors.

The verbal abuse of women seems to have been given sanction and permission. False rumors, lies and unsubstantiated news vignettes are treated by millions of Americans without question, debate or investigation, as if these things alone are the voice of absolute truth. Around the world we see the widespread obliteration of wildlife, religious bigotry between Buddhists and Muslims in Myanmar, violence against women and even slavery still upheld. Everywhere around the world we see that millions of human beings if not billions, have allowed their passionate emotions and brute aggression to get the better of them, something that intelligent robots, fortunately, have not learned to imitate, at least not yet. The crucial point is that frequently these moral ills are spread or they are helped to spread like disease epidemics, by new technologies one abuses.

Today some people are fearful that robots and artificial intelligence one day will wipe out the human race. Based on human behavior however, it is much more likely that if history is any judge, the human species will obliterate itself possibly within one or two decades, such as perhaps by war and thermonuclear weapons, a worldwide Internet malware attack, serious disease pandemics, environmental degradation worldwide or some combination of all these potential disasters, long before Homo sapiens gets supplanted by intelligent machines. The ways that humans use their new technologies says much more about Homo sapiens than it does about artificial intelligence and robots.

Just how "sapiens" is Homo sapiens?

First gunpowder was developed then dynamite and hand grenades. First came the airplane with the work of the Wright brothers and nuclear fission in the German laboratory of Otto Hahn, then fighter bombers and thermonuclear weapons. As human beings who either are citizens of the United States or citizens somewhere else in the world, we comprise today both a nation and a world civilization in serious trouble with ourselves. Now at this stage in our technological development just below Type I on the Kardeshev scale, we confront artificial intelligence and robotics, still newer technologies which, although certainly not evil in themselves anymore than are rainstorms, galaxies and grizzlies, could provide humans with new means to behave very badly as an "intelligent" species.

In his book *Cosmos*, planetary astronomer Carl Sagan discussed how a planet that has a world civilization reaches a very crucial point in its history, a time for which their future survival toward becoming a Type I civilization on the Kardeshev scale depends almost exclusively upon what mistakes they make with their new technological discoveries.

If anyone does not think that collective humanity is on the brink right now of destroying itself along with this planet's biosphere due to the ways in which human beings repeatedly use new technologies to uphold and to act out their destructive patterns of thought and behavior, that person either is deluded or else a very poor student of human history. Therefore if any reader is astute enough or wise enough to see the possible future dangers that lie in wait from the use of robots and AI by humans, they will find arguments in Part II that are relevant to the matter. But as one reads the last three chapters one ought to understand that this book is *not* a polemic against any further advancement of artificial intelligence and robotics. For instance it makes no sense to argue one should rid the entire world of automobiles because some irresponsible human beings do drive drunk or under the influence of mind altering drugs. Rather the last three chapters only are a reminder that frequently humans have abused new technologies in the past (we shall take a look at some historical precedents as well as some modern day examples) and even today human beings still tend to use new technologies in immoral and in irresponsible ways.

Perhaps you might not agree with some of the conclusions of this author that are expressed in the last three chapters, thinking it is too polemical in tone. Of course everyone is entitled to some opinion, but if you were taking a brisk walk one night through unfamiliar woods and you were about to fall over the ledge of a cliff to your death because you became distracted, which would you prefer? Would you prefer someone nearby to say in soft and guarded tones: "I think it might be advisable for you to watch where you are going," or for that person to yell at you "Look out!" before you took your final step? Believe it or not we are at a time when world civilization as it stands today could be on the brink of destruction. The last three chapters are a warning shouted in the darkness.

Still one must agree at the least, that intelligent robots either can be used for good or for evil ends by men and women, just as a laser can be used for healing in surgery and a hydrogen bomb can be used as a weapon of war.

Part I

Robots in Mind

Chapter 1

Automatons in Western Antiquity

Myths from ancient Greece are replete with examples of automatons of many different types. The giant eagle that tore at the liver of Prometheus everyday in *Prometheus Bound* (written by Aeschylus, 525 BCE-456 BCE), who was condemned by the jealous God Zeus to be bound by chains and imprisoned in the Caucasus Mountains, was in a sense an automaton. The account by Ovid in *The Metamorphosis*, on the statue of a flawless female crafted by Pygmalion who fell in love with it as it was for him the perfect "ideal woman," is known by many who are familiar with ancient Greek literature. Down through the centuries as the tale was retold the perfect female statue in Ovid's account eventually was given the name Galatea. Even the story underwent various metamorphoses of sorts, first in the 1913 play by George Bernard Shaw named *Pygmalion*, where the marble statue is replaced with a spirited, vibrant but illiterate Cockney English flower girl named Eliza Doolittle (the name "Eliza" figures prominently in the early years of artificial intelligence research, as we shall see in this book), then in the 1956 Broadway musical production *My Fair Lady* by Lerner and Lowe[1].

Another more definitive example of an automaton in ancient literature is in the *Argonautica* by Apollonius of Rhodes (3rd century, BCE). After his return from Colchis (located in modern day Georgia on the coast of the Black Sea), having acquired the Golden Fleece for King Pelias of Lolcos, Jason sails onward aboard the *Argo* toward the island of Crete. As they approach the shore the sailors are attacked by Talos who tosses boulders at Jason and his men. The giant automaton was constructed of bronze, supposedly by the god Hephaestos (the ancient Greek prototype of the Roman god Vulcan) to protect Europa (See the Wiki link for Talos in the Preface). Apollonius describes him as follows:

> And Talos, the man of bronze, as he broke off rocks from the hard cliff, stayed them from fastening hawsers to the shore, when they came to the roadstead of Dicte's haven. He was of the stock of bronze, of the men sprung from ash-trees, the last left among the sons of the gods; and the son of Cronos gave him to Europa to be the warder of Crete and to stride round the island thrice a day with his feet of bronze. Now in all the rest of his body and limbs was he fashioned of bronze and invulnerable; but beneath the sinew by his ankle was a blood-red vein; and this, with its issues of life and death, was covered by a thin skin. So the heroes, though outworn with toil, quickly backed their ship from the land in sore dismay.

---

[1] In 1964 the theatrical production became a major motion picture musical, with Rex Harrison as Professor Henry Higgins ("Pygmalion") and Miss Audrey Hepburn as Eliza ("Galatea")

Jason and his men are saved by the sorceress Medea however, who uses her powers for the giant bronze robot's undoing. Due to the spell cast by Medea, Talos succumbs to the *Keres* sisters, evil spirits of death and doom. The powerful giant suddenly dislodges the nail in his foot that keeps the precious divine ichor flowing in his vein and then collapses as he bleeds to death:

> So Talos, for all his frame of bronze, yielded the victory to the might of Medea the sorceress. And as he was heaving massy rocks to stay them from reaching the haven, he grazed his ankle on a pointed crag; and the ichor gushed forth like melted lead; and not long thereafter did he stand towering on the jutting cliff. But even as some huge pine, high up on the mountains, which woodmen have left half hewn through by their sharp axes when they returned from the forest–at first it shivers in the wind by night, then at last snaps at the stump and crashes down; so Talos for a while stood on his tireless feet, swaying to and fro, when at last, all strengthless, fell with a mighty thud.

Actually contrary to being an epic poem about bronze automatons, the *Argonautica* tells in part a love story about Jason and Medea! Unfortunately this romance between the famous argonaut and the powerful witch and princess from Colchis comes to a violent and tragic end within the pages of *Medea*, which was written by the Greek tragedian Euripides (480 BCE - 406 BCE) around 431 BCE. The brief account about Talos though, does indicate to us that ancient writers and poets along with their readers, were familiar with the concept of mechanical facsimiles of animals and human beings.

1.1   Mechanical Automatons of the *Ancien Régime*

1.1.1   The Versailles Gardens of the "Sun King," René Descartes and Francine

Natural philosophers, anatomists and engineers in seventeenth century Europe had been influenced extensively by the enormously powerful work of Sir Isaac Newton. The *Principia Mathematica* was the bright light that would lead the way to new discoveries in physics and developments in mechanical engineering many thought could be explained by determinism alone. Back then some people believed there was no event in nature physical or human that could not be explained by deterministic physical laws of nature that had been formulated into the elegant system devised by Newton. The Moon locked in its orbit around Earth and the Earth itself restricted in its motion about the Sun, the way a bridge supported the weight of a leaf, a horse or an oxcart, everything was subject to physical cause and physical effect and all of this was predictable completely if one knew all the variables involved to study the mechanistic system at hand, be it a distant star, a solar system or a road or bridge under construction. So if Newtonian physics can explain how the Earth orbits the Sun and how a bridge can support an oxcart, why can it not explain the motion of a human arm or finger, since it can explain the motion of a pulley or a lever?

At least that is how the world seemed to some people at the time, people who had delved into the new wonders and possibilities of Newtonian mechanics.

Fortunately the philosopher René Descartes did not fall entirely into this restrictive mode of deterministic thinking. As was pointed out by author James P. Hogan in his book, *Mind Matters: Exploring the World of Artificial Intelligence*, Descartes was able to divide the world into two dualistic opposites, *res cogitans* and *res extensa*. There are events determined by human will. There also are other events, determined by physical cause and effect.

Still anatomists, engineers and others during the reign of Louis XIV maintained the idea that the whole universe ran like some kind of brilliantly manufactured Swiss clock. If it is wound up well by some Master Watchmaker the cosmos will tick on forever without error. Some inventors wished to demonstrate their own talent and flair for developing mechanistic creations. As Hogan points out in his book, the Royal Gardens at Versailles displayed their automated robot inventions to many a surprised or astonished human visitor: Hydraulically powered automatons of men, women and gods from Greek mythology whizzed, darted and glided about the Gardens at Versailles to entertain and to marvel human visitors, to make fascinating sounds and to perform on musical instruments.

Descartes himself was not amiss in the invention of automatons. One female looking automaton he had built was named Francine. This automaton was so realistic in its physical appearance and so convincing in its automated movements that a ship captain who had seen the female automaton tossed it overboard when Descartes was on a voyage.

1.1.2   An Android Dulcimer Player for Marie Antoinette

The dulcimer[2] is a string instrument that migrated its way first from the Middle East as far away as medieval Persia then to Europe during the Crusades and then finally eventually even to the United States where it has found popularity among some Appalachian musicians. It was constructed into various forms but in essence it looked like a wooden box shaped like a trapezoid. The performer either would pluck the strings or strike them one at a time gently with two small hammers or mallets. An eighteenth century German musician named Pantaleon Hebenstreit built a dulcimer large enough so that it resembled a clavichord or harpsichord. The player would sit at the pantaleon and perform by striking the strings gently with two mallets. As the notes were struck in succession the tones would blend together since there was no damping of the strings as there is with the damping pedals on the modern piano.

The dulcimer has a pure, ethereal sound to it. One might be tempted to suggest its sound is reminiscent of the "music of the spheres" description one has found in various forms of literature. Anyone who has heard the zither music performed in the movie *The Third Man* (with actors Joseph Cotton, Orson Wells and British actor Trevor Howard), will get some sense of the sound of the dulcimer, only unlike the zither which sounds deep and mellow at least in a subjective sense, the tones on the dulcimer are lighter and more faint. Perhaps a better comparison would be to the glass armonica invented by Benjamin Franklin.

In 1772 musician Pierre Kintzing and cabinetmaker David Roentgen presented Queen Marie Antoinette with an extraordinary gift. It was a dulcimer that looked actually like the pantaleon, one at which the player sat before to strike the strings with the two dainty hammers in order to play a melody.

Remarkably though what also was different about this dulcimer presented to the ill fated Bourbon queen of France, was that no human performer played the instrument. The performer, or *Joueuse* in French, was an elegant looking female automaton. *La*

---

[2]The word is made from two different foreign words, namely "dulce" from Latin meaning "sweet," and "melos" from the ancient Greek word meaning "melody."

Joueuse de Tympanon, or the (female) dulcimer player became quite a sensation at the royal court. Unfortunately during and after the tumult and uproar of the French Revolution the *Joueuse de Tympanon* fell out of fashion among a devastated nobility.

At the time of this writing the dulcimer performing automaton, or *Joueuse de Tympanon* can be observed in actual performance at www.youtube.com.

1.1.3   Does it really, really play Chess?

In the late eighteenth century, as Colonial North America reeled from the French and Indian War, an automaton appeared in Europe that came to be called the mechanical "Schachtürke" or mechanical "Turkish chess player," a contraption that was not formed by a god like Hephaestos but designed actually by a German inventor named Wolfgang von Kempelen (1734-1804). The mechanical Turkish chessplayer had the upper torso of a man with two arms and hands. One hand held in it a long Turkish smoking pipe. The mechanical man was dressed in ornate Oriental garb and turban and was seated at a chess cabinet on top of which one saw a chessboard. Von Kempelen built the automaton to impress the Empress Maria Theresa. When animated the automaton appeared to be able to defeat many a human chess master at the game, even the likes of Napoleon Bonaparte.

In truth it only was a clever mechanical ruse by von Kempelen. A human operator who knew chess hid inside the cabinet. By moving various hidden levers, he was able to get the automaton to make the right chess moves and in so doing, to astound many a human opponent. Von Kempelen had better success with his Speaking Machine. This was a mechanical speech synthesizer he constructed that consisted of a bellows and a reed, both attached to a kind of voice box with holes in the top. With some other rubber attachments von Kempelen was able to generate audible sounds from his device that sounded like various vowel and consonant pronunciation patterns, to synthesize specific words known from various spoken human languages.

Whether he realized it or not von Kempelen with his *Schachtürke* and Speaking Machine demonstrated a knowledge of ideas that would become more advanced more than two hundred thirty years after his time such as teleoperated robots and computers that can play checkers, chess and Go with humans, as we shall see in Part II.

Chapter 2

Fantasy, Horror and Science Fiction

It is said that necessity is the mother of invention. It also can be said that no new scientific theory or groundbreaking mathematical proof, physical robot or even orchestral symphony for that matter, can emerge without the initial seed of an idea. As a youth and before he began to work on his Special Theory of Relativity, Albert Einstein imagined what it would be like if someone could ride a beam of light. In the nineties at Princeton University mathematician Andrew Wiles had various ideas about elliptic curves, modular functions and modular forms[1] in his mind when he began to work on a proof[2] to Fermat's Last Theorem. More than likely the robotics engineers at Boston Dynamics were thinking about how pack animals manage heavy cargo on their backs, when they were conceiving the initial designs for Big Dog. Ludwig van Beethoven (1770–1827) was thinking about a new invention called the metronome before he completed his Eighth Symphony.

Frequently an artist, painter or science fiction novelist conceives the unique idea long before the inventor shapes that idea later into reality. Before Sputnik and Telstar there was Arthur Clarke with his ideas about manmade satellites.

Today the computable universe hypothesis in physics asserts that everything in the universe derives from computation. Did the universe "compute" all that there is within it, out from the quantum vacuum thirteen billion years ago, including all light quanta, protons, neutrinos, galaxies, stars, fire hydrants and newspapers? One is reminded of the ideas Spinoza had with regard to *natura naturans* and *natura naturata*.

It seems that in an analogous way the actual physical robots and computers we see in the world today or that might be designed and built in future years, have been anticipated already within works of Gothic horror, science fiction and fantasy, works conceived within the minds of some imaginative human writers in the past.

We now shall explore some examples of these in fiction and in cinema.

---

[1]A "modular function" is a certain kind of function defined on the complex plane, that has certain invariant "conformal transformations." A "modular form" is a very specific kind of complex function that has both certain invariant conformal transformations as well as values everywhere on the complex plane, even at "points at infinity," such as in a one dimensional projective space for complex numbers on a curve.

[2]A good source for laypeople on Wiles's extraordinary result can be found in the Simon Singh book *Fermat's Enigma: The Epic Quest to solve the World's greatest Mathematical Problem* (See the Bibliography).

2.1 Frankenstein

> Remember, that I am thy creature; I ought to be called thy Adam; but I am
> rather the fallen angel, whom thou drivest from joy for no misdeed.

The work *Frankenstein: Or, the Modern Prometheus* has enthralled millions of readers worldwide and has proven itself to be a classic of Gothic horror fiction that also inspired the famous 1931 movie version in which British actor Boris Karloff portrayed the Monster while Colin Clive donned the role of Viktor Frankenstein. In 1818 Mary Wollstonecraft Shelley (1797–1851) had published the first edition of her novel. Today the theme of the story is among the most famous clichés within the canon of rules for science fiction and horror screenwriters: Mankind should leave the creation of life to God.

We give a very brief synopsis here for this epistolary novel, only because this Gothic tale is so familiar that most of the details in the plot are well known. Viktor Frankenstein is an exceptionally gifted researcher who completes his studies in anatomy, chemistry and natural philosophy. Then one day by patching together, one only can assume, various parts of cadavers and organic tissue gathered from "the dissecting room and the slaughterhouse," Viktor uses some mysterious elemental force of nature (that might or might not be related to studies in galvanism) to convey his creation across the boundaries of his new but apparently apocryphal science, from the state of inanimate organic matter to life.

At once Viktor is horrified and frightened, upon seeing the yellow and watery eye of his actual creation in the light of a candle. He describes the creature as "hideous," as a "miserable wretch," something of which the author of *La Divina Commedia* "could not have conceived." Viktor flees in horror from his creation and seeks solace within the arms of his beloved and beautiful sweetheart Elizabeth Lavenza.

After suffering the humiliation of revulsion and rejection by those humans who come into contact with it, the Monster rebukes Viktor for creating it. At first it describes itself as being Viktor's Adam just as Adam was God's creature in the Biblical Genesis, but then compares itself to Satan. It pleads for Viktor to create for it a female version of itself. At first Viktor complies with the request, but then he contemplates the significance of such an act. Since it could mean the creation of a hideous and evil race of creatures Viktor destroys the female creature. The Monster is enraged and goes on a killing spree. He kills those close to Viktor, meaning Elizabeth, Viktor's brother William and Viktor's close friend John Clerval.

Mary Wollstonecraft's *Frankenstein* is a literary time capsule of sorts for nineteenth century Romanticism. As such it raises many questions. In the novel Viktor makes it clear he labored on his creation for two years. One can ask how then, knowing from where he obtained the parts for his creation, could he be so horrified at the final result when viewing it in the light of a dim candle? Viktor's horror at the sight of the creature puts him into extreme and altered states of emotionalism: Terror, revulsion, disgust, fear, guilt all in the true and unabashed spirit of Gothic fiction and romance. Those humans who encounter the creature in its travels experience similar states of heightened emotions. The novel gives the impression that Elizabeth is good and morally pure because she is physically beautiful and that the Monster is evil because it is physically repulsive and hideous. Perhaps this is one of the failings of Romanticism, in that the meaning of science, the world, beauty and repulsiveness, good and evil all are identified and categorized within the context of overgeneralizations and heightened human emotive responses, not through any recourse to logic or rational thought.

Obviously Viktor Frankenstein had lost all control of his creation, moral or otherwise. Actually despite all his revulsion at the hideous "wretch," ironically he was the one who created his own evil monster. That being the case, who should have answered for the evil actions of the creature, the creature itself, or its creator?

Could the same thing or something similar or even worse happen within the next one hundred years, if some future researcher in robotics and artificial intelligence gives superior artificial intelligence and possibly even simulated emotional responses to a machine, but does not take care to build the machine so that it does not exhibit the very worst aspects of human behavior? Is it even a good idea to build intelligent robots that behave in the emotional ways that we humans behave? We explore these questions in Chapter 13 and Chapter 14.

We mention briefly about the other story to which Mary Wollstonecraft Shelley alluded, namely the story of Prometheus, one of the Titans who gave fire to mortals, due to an ongoing rivalry between Prometheus and Zeus. The tale is given in the ancient Greek tragedy *Prometheus Bound* by Aeschylus (523 BCE–456 BCE), although some scholars today doubt the authorship. Zeus punishes Prometheus by having him chained to a rock on a remote island where a giant eagle feasts daily on the Titan's liver for eternity. Unlike in *Frankenstein* the conflict derives from the feud between the Titan and the All Father God, not from the actions of humans or monsters created in the laboratory, as is the respective cases in the Biblical Genesis and in *Frankenstein*.

2.2 The Golem

The Golem is a legendary creature that, in stark contrast to being the product of vivisection, medicine or engineering, evolved actually from the motley milieu of Jewish medieval life, religion, folklore, superstition and culture within the city of Prague. We mention it here mainly because, although the Golem supposedly was activated not through science but by the practice of superstition or religion, still it was a robot in the definitive sense of the word in that it was more or less an unthinking slave that was a true automaton, sort of like an old fashioned wind up mechanical toy that walks about mindlessly until finally the spring exhausts its resource of kinetic and potential energy. Legend has it that this creature was fashioned from river clay and animated by inscribing upon its forehead one of the names of God as this name is defined in the Old Testament, the Torah and in Talmudic writings. There are several different versions of the story however. Among the most famous is the 1914 German novel *Der Golem* (*The Golem*) by Gustav Meyrink, and the 1920 film *Der Golem, wie er in die Welt kam* (*The Golem, how He came into the World*).

The Meyrink novel is a disappointment for those readers who expect to be captivated or spellbound by the antics of some superhuman automaton made from clay. The Meyrink novel to the contrary focuses rather on the mental anguish of its first person narrator and so the novel is expressionist in its style and tone. This holds throughout the novel as the protagonist, who identifies himself as being someone named Athanasius Pernath, interacts with various people and places in the Jewish ghetto of Prague. The reader is left to ponder whether the Golem merely only is some mysterious mental hobgoblin conjured up from Pernath's feverish state of mind or some kind of nebulous allegorical figure standing in for the people and places in the tale. Expressionism had developed and metamorphosed into various artistic forms after the beginning of the twentieth century with its first world war in history and mechanized weaponry and was typified in art, in film and in drama, such as in the painting *The Scream* by Edvard Munch (1863–1944), *Das Kabinett des Doktor Caligari* (*The Cabinet of Doctor Caligari*. This film also is at Archive.org), and as we already have seen in *The Golem, how He came into the World* (1920s) and in the play *The Emperor Jones*, by American playwright Eugene O'Neal (1888–1953).

In contrast to the Meyrink novel the 1920 film (See the Wiki link for the film in the Preface) is a silent film *tour de force*[3], if the viewer wants to see giant, rampaging and mindless clay automatons brought to life by astrologically related incantations and evil spirits. In the beginning of the film after indulging in certain astrological rituals and making related prognostications, a rabbi named Loew brings the clay form to life through supernatural means that involves activating it by placing an amulet into its clay chest, in order to help the community of Jews who have been ordered to leave the city by the Emperor. At first the rabbi uses the Golem to perform menial chores but when Rabbi Loew uses the Golem (played in the film by the German actor and director Paul Wegener) to prevent the palace from collapsing and killing the Emperor and his guests, the Jewish inhabitants of the Prague ghetto are spared from the expulsion decree.

Eventually though things go awry. Loew orders the automaton to remove an unwanted suitor of his daughter Miriam from a building. But after the evil spirit Astaroth manipulates the clay automaton the Golem tosses the suitor, a knight named Florian, off the roof. To make matters worse the Golem drags the horrified Miriam through the streets using the young woman's own long hair as some sort of dragging rope. Finally the creature's mindless rampage is brought to an abrupt end when a mere child removes the charm from its chest.

If one likes to watch monster robots rampaging inside horror films without a lot of computer generated imagery it does not get much better than this! Even though in *Der Golem* the emphasis is on superstition and religion the film does show what can happen when some kind of powerful automated system gets out of control, in particular when there is not much of a mind behind its actions.

2.3   *Tik-Tok*

There is no dearth of legends and stories about steam powered, mechanical and electrically powered bipedal automatons who ran on two legs at high speed across the Western American prairie like The Steam Man in *The Huge Hunter, or the Steam Man of the Prairies* (a novel by Edward S. Ellis), or shared adventures alongside Pancho Villa like the legendary Boilerplate. Yet one often neglected mechanical man in fiction was Tik-Tok, an automaton devised by L. Frank Baum in his Oz stories. Yet back in 1959 I never had heard of Tik-Tok.

This author can remember watching *The Wizard of Oz* as early as 1959, two years after the Sputnik launch (which in fact was the first machine to orbit Earth), back at a time when the learning of English grammar in most American public schools, at least in the northeast, began with *Dick and Jane* books and was solidified later with *The Weekly Reader*. As a child my favorite Baum character was The Scarecrow, although the Tin Man also had his own special appeal for me.

---

[3]This author watched the film at Archive.org and can guarantee this silent movie is pretty atmospheric and exciting for those who like to watch old vintage films.

Although he was a famous and popular writer of stories for children, L. Frank Baum had a very unflattering and controversial side to his human nature. Yet even so it goes without saying that Baum had built up a huge and compelling mythology of Oz related characters. The Tin Man was not the only mechanized character in his collection of stories. Another of Baum's mechanical characters was a copper mechanical man named Tik-Tok[4], something perhaps far more reminiscent of what is a robot. It walked, talked and thought all through the power given to it when its corresponding springs were wound up. Three Baum Oz characters named Shaggy, Betsy and Polychrome wind up the mechanical man and then they all interact with it:

> So Betsy wound him under his left arm, and at once little flashes of light began to show in the top of his head, which was proof that he had begun to think.
>
> "Now, then," said Shaggy, "wind up his phonograph."
>
> "What's that?" she asked.
>
> "Why, his talking-machine. His thoughts may be interesting, but they don't tell us anything."
>
> So Betsy wound the copper man under his right arm, and then from the interior of his copper body came in jerky tones the words: "Ma-ny thanks!"
>
> "Hurrah!" cried Shaggy, joyfully, and he slapped Tik-Tok upon the back in such a hearty manner that the copper man lost his balance and tumbled to the ground in a heap. But the clockwork that enabled him to speak had been wound up and he kept saying: "Pick-me-up! Pick-me-up! Pick-me-up!" until they had again raised him and balanced him upon his feet, when he added politely: "Ma-ny thanks!"

Tik-Tok (See the Wiki link for Tik-Tok in the Preface) is a true robot and automaton, at least in literature. He does not succumb to the overwhelming urges of violent emotion as does the Frankenstein monster, nor is it the product of supernatural revelation as is the Golem. One winds him up and he walks, talks, obeys commands. When someone whips him as does occur in the story he does not bleed or cry or become vengeful. Actually inflicting Tik-Tok with violence says a lot more about the abuser than it does about the machine man.

2.4 *Karel Čapek's R. U. R.*

When the play *R. U. R.* (stands for *Rossum's Universal Robots*) by Karel Čapek (1890–1938) was premiered it caused quite a sensation when it introduced the word *robot* to the theater going public (See the Wiki link for R. U. R., in the Preface). It also motivated a worldwide interest in robots. The theme of this very famous Czech science fiction play is similar at least in some ways to the point made in the 1925 film *Metropolis* by Fritz Lang. If one day the human race does create a race of intelligent robots or androids, it should be prepared to confront very serious moral and ethical issues, possibly even violent revolt by humanity's creations.

On a remote island a man named Rossum (which might be related to the Czech word

---

[4]The Tik-Tok character appears in fact on the book cover for this book

rozum for "reason"), accompanied by his nephew, creates a race of intelligent artificial humans, composed of a synthetic material that resembles protoplasm. Rossum's nephew exploits the new beings for commercial purposes and soon they are assembled part by part in mass production similar to how American industrialist Henry Ford had Model T's assembled at his Michigan plant. Eventually a human rights organization led by a woman named Helena Glory begins to challenge the morality of the mass production of the robots and demands their liberation. The robots themselves however begin to resent their lower status, menial positions and lack of rights as the years progress, so in time they attack Rossum's factory and rebel against the human race, which falls victim to mass extermination at the hands of the robots.

But one of Rossum's human corporate officers, the Clerk of the Works named Alquist, has survived the genocide. He has dedicated himself to improving the robot design but he needs an experimental formula for the design and the original design was destroyed by a robot named Helena (this Helena is not to be confused with the human female named Helena Glory). The robots permit the human Alquist, the last human remaining, to dissect robots in his experiments to make the improvements. At the end of the play not only have the humans been exterminated due to their own material greed and callousness, but the robots themselves seemed to have been corrupted as were the humans, in that their demands for technological self-improvement makes them develop a callous attitude toward any robots among them who might get dissected for the cause. Like father, like son, apparently, would not apply alone to human children.

Some computer scientists, artificial intelligence researchers and roboticists in the United States and in Japan already have designed and built android machines that resemble humans physically with remarkable aplomb, such as the androids that were displayed at the National Museum for Emerging Science in Tokyo. But if future machines that look too human are made too intelligent as well, it could one day cause machines that resemble humans to emulate not only the very best of human behavior but also the very worst of human behavior, as Karel Čapek showed in his drama. At present some lay people already have expressed interest in the development of sex robots. Isaac Asimov even explored the idea of sex between humans and robots. But if humans today can engage in morally unacceptable, even pathological behavior, is it a good thing for machines to emulate that behavior through learning? For example it is horrid enough that humans today do engage in violence, sexual exploitation, murder and sexual assault. It would be a bizarre world indeed if intelligent android machines begin to run algorithms that make them do the same kinds of things.

2.5 Robots, Empire, Isaac Asimov and The Three Laws of Robotics

Perhaps no other science fiction author has written more extensively and persistently about robots and androids within his short stories, anthologies and novels than has Isaac Asimov. Within the famous SF anthology *I, Robot* it was Asimov who defined The Three Laws of Robotics, although various science fiction authors since have defined their own versions of these laws.

The Caves of Steel and *The Naked Sun* are science fiction stories about robots, but both actually are cross genre works that combine science fiction with murder mysteries. In the first novel New York homicide detective Elijah Baley is assigned the task of finding out who murdered a human ambassador from one of the "Spacer" worlds. The victim was an advocate in the cause of robots and robotics. R. Daneel Olivaw, which assists Baley in his investigation, is a robot that strongly resembles a human. Baley's New York of the future has become part of a closed environmental system illuminated by artificial lighting instead of by stars and sunlight. Asimov treats the possibility that humans in such a civilization sealed off from nature might be the perfect place for humans to become agoraphobic, as is the case with Elijah Baley. The human and robot detective partners appear again in the novel's sequel *The Naked Sun*, when they must solve the murder of a Spacer scientist who was a researcher in a science that is reminiscent of in vitro fertilization. In *The Naked Sun* Baley, a married New Yorker, meets his future Solarian mistress, Gladia, a character who turns up in several later novels by Asimov, such as in *Robots and Empire*.

Asimov's Earth and the Spacer worlds are a study in contrasts. The Earthers have restricted robot development and the human species has overpopulated the planet. In contrast the human inhabitants on the Spacer worlds elsewhere in the galaxy have built so many robots they outnumber the humans. In addition the Spacer worlds are underpopulated and humans there detest physical contact. These stark differences lead to distrust and outright hostility between the two diametrically opposite societies.

In *I, Robot* Asimov defines The Three Laws of Robotics as follows:

1. A robot may not injure a human being or, through inaction, allow a human being to come to harm.

2. A robot must obey the orders given it by human beings, except where such orders would conflict with the First Law.

3. A robot must protect its own existence as long as such protection does not conflict with the First or Second Laws.

In Chapter 18 in his novel *Robots and Empire*, Asimov has his robot character R. Daneel Olivaw define an additional Zero-th Law:

0. A robot may not harm humanity, or, by inaction, allow humanity to come to harm.

Two robots named R. Daneel Olivaw and R. Giskard Reventlov, both the inventions of roboticist Doctor Hans Fastolfe, must stop two roboticists, Kelden Amadiro and Levular Mandamus, from destroying the Earth to prevent humanity from populating the Galaxy. Daneel devises The Zeroth Law of Robotics to stop them.

Robots and Empire is one novel that in a sense is a part of a broad and elaborate tapestry of novels and ideas all by Asimov, to combine concepts like telepathic robots, robots having sex with humans as mechanized paramours, "psychohistory," human migration across the Galaxy and even murder mysteries. What I wish to focus on here though is that in both novels *The Robots of Dawn* and its sequel *Robots and Empire*, Daneel and Giskard demonstrate an incredible ability to make extremely complex decisions that would challenge the most clever human. In order to devise a new law of robotics, did a highly intelligent robot like Daneel have to design new algorithms then run them with programs to make the decision? What kind of algorithmic processing did Giskard have to implement to reach the decision to use telepathy on a human villain? In the future what kinds of intelligent machines might have to make very complex decisions, and will they do it well?

2.6   Fritz Lang's *Metropolis*

The robot Doppelgängerin that replaces the human character Maria does play a pivotal role in the overall story behind Fritz Lang's *Metropolis* (1927). But the main plot and theme that drive this silent film of science fiction has to do more with the social conflict, economic disparity and differences between the lives of the wealthy human elite and the lives of underpaid human industrial workers inside the fictional city of Metropolis.

This film, with its screenplay written by Lang (1890–1976) and his wife Thea von Harbou (1888–1954), probably is one of Fritz Lang's best known cinematic masterpieces[5], made all the more compelling by Lang's brilliant direction and in particular due to the film's quasi Wagnerian music score composed by Gottfried Huppertz, the initial gripping montage shots of the giant cogs, shafts and gears that drive the city from its lower depths and that are kept in operation day and night by mechanized and exhausted human workers, the diametrically opposed characterizations for the saintly humanitarian Maria and her propagandizing, lust inciting Doppelgängerin (both played by German actress Brigitte Helm) and the stark, visual contrasts between the pleasure and leisure of the city's upper class with the suffering and toil of the workers.

In the film the scenes that contrast the highrise towers of the upper elite with the underground lairs of the poor workers reveals a modernistic dystopia, in which wealth and technology are reserved for the rich while the lives of the poor do not count for much in value except in terms of manual labor.

In an historical context Lang's film is a basket of ironies. Some years after the film made its debut Lang's wife von Harbou, who collaborated with him to write the screenplay about the human oppression of the poor by the rich, joined Germany's National Socialist Party. After 1933 in Germany the leaders like Adolph Hitler and Paul Josef Göebbels did appear at the beginning to eliminate classism and social distinctions from German society but at a horrific cost in terms of human life. In early 1933 Doctor Göebbels had met with some major German film producers and executives at Berlin's Hotel Kaiserhof. The claim has been made that Lang also was present at this meeting. In any event shortly after this meeting with Nazi Germany's new Propaganda Minister, Fritz Lang left Germany for the USA.

In *Metropolis* the stadium in the Club of the Sons seems to be a cinematic metaphor for the Nazi tyranny, representative of the gargantuan but monolithic grandiosity in building design, architecture and construction, building designs for which Hitler was so fond, the type of grandiosity that was put on display by Leni Riefenstahl in her film *Triumph des Willens*[6]. Another more tragic irony was that one of the opening scenes in *Metropolis* shows two ranks of human workers, one exhausted and the other slightly more animated, marching very slowly in opposite directions like robots into and out from a huge elevator just at the start of another work shift. What is ironic about this is that

---

[5]One of his other cinematic masterpieces was the German film *M*, which starred actor Peter Lorre as a psychopathic child murderer in this somewhat expressionistic German film. One of Lang's successful American films was *Fury*, with Spencer Tracy and Sylvia Sydney.

[6]*Triumph of the Will.*

there are photographs that show inmates in Dachau concentration camp marching in a similar way to labor in a stone quarry. One can imagine that Europe's slave laborers also marched this way at the beginning of their work shifts, as they were forced to construct the rocket research facilities for the Luftwaffe at Peenemünde.

What truly is notable and memorable about the Doppelgängerin or ersatz Maria in *Metropolis* was that the robot not only was a perfect copy of Maria's outer physical appearance. It was highly intelligent and quite capable of deception, easily fooling young men into believing it was the beautiful female human named Maria. It is claimed sometimes that "Beauty is only skin deep." From the success of the robot's efforts it would appear that when the need for deception of some humans arises somewhere, sometimes physical beauty is all that is required! In fact the idea of a robot android deceiving humans because humans take for granted it too is human has cropped up in many other works of science fiction, such as in the movie *Alien* (1979) by director Ridley Scott. In that film a spaceship's science officer actually is an android not a human, who uses that fact to entice the human crew members to allow an injured crew member to board the ship *Nostromo* even though the injured man has become both a host and prey to an evil and predatory alien parasite.

Who or what, probably, could benefit from the use of a highly intelligent and engineered android that resembles a human to the point of it being capable of the complete deception of other human beings who interact with it? Such an invention might get its inventor the Turing Award! There is no dearth of villains one can list who could benefit from such remarkable technology. Unscrupulous corporate executives could design androids to resemble and to behave exactly as do certain corporate officers on the rival corporate board of a competitor enterprise, using these androids as industrial spies. Organized gangs of criminals would have perfect assassins available to infiltrate rival gangs. National governments could design an android that looks exactly like someone inside some enemy country, for example taking the place of a human domestic servant or human services worker in some legislative building. Technology after all, even since the very beginning of the Industrial Revolution and including such things as the cotton gin, dynamite and the airplane, has been used for evil purposes as well as for the performance of good deeds. Yet in defense of science and technology we humans should be reminded of the fact that the evil uses to which technology frequently is put does not lie itself within the actual invention.

2.7 The Day the Earth Stood Still

Klaatu, barada, nikto.

More than likely "Klaatu barada nikto" in *The Day the Earth Stood Still* (1951) by film director Robert Wise is the most famous alien language phrase from cinematic science fiction and the eight foot tall robot Gort, called Gnut in the short story *Farewell to the Master* by Harry Bates, still remains from the Truman era as one of the most imposing robots on the movie screen. Gort's metallic frame was supposed to be made from some unknown alien alloy. According to the alien character Klaatu (played by British actor Michael Rennie) the robot was capable of destroying the Earth. Klaatu states his planet is two hundred fifty million miles away from planet Earth. Given that assumption the planet Mars then would have to be closer to Earth than is Klaatu's world. But the planet Jupiter is much further away from Earth than is Klaatu's world (See the Wiki link for

The Asteroid Belt in the Preface).

The asteroid belt lies between the orbits for Mars and Jupiter. Think of a graph where the horizontal axis is for the same distance in astronomical units that asteroids are from the Sun while the vertical axis is the number of asteroids that have the same AU distance away from the Sun. When one considers the frequency spread for all the asteroids considered in the asteroid belt, the result is a series of histograms that resemble side by side Gaussian distribution plots, where the values for the random variable are given by the AU distances along the horizontal axis. The points that show an extremely low number of asteroids for an AU distance are located at what are called "Kirkwood gaps," named after astronomer Daniel Kirkwood (1814–1895). Asteroids in the Kirkwood gaps have very chaotic, perturbed orbits. Any asteroids in the graphs that lie within the adjacent frequency spreads between the gaps, have more or less stable orbits (See the Wiki link for "Kirkwood Gaps" in the Preface).

The asteroids within the 3:2 gap range have orbits perturbed by Jupiter. They complete three periodic orbits to every two periodic orbits of Jupiter by Kepler's Third Law, due to "orbital resonance," or chaotic motions caused by gravitational perturbations on the small asteroids in this gap by the much larger, gas giant planet Jupiter. Recall from the 1951 version of the movie that Klaatu claimed his planet was two hundred million miles away from Earth? Since Jupiter is about 5.2 astronomical units away from the Sun and with Mars a distance of 1.52 AUs, that distance of two hundred fifty million miles away from Earth would put Klaatu's world somewhere inside the asteroid belt between Mars and Jupiter.

Suppose both Klaatu's world and Earth are aligned in conjunction, that is, on the same side of the Sun as viewed from his world. One then can use Kepler's Third Law to show that, since Klaatu's planet would have to lie somewhere between the planetary orbits of Mars and Jupiter, its orbital period would have to be somewhere between the orbital period for Mars and the orbital period for Jupiter. If we estimate the distance Klaatu's planet has from the Sun to be roughly equal to be the sum of the distance of the Earth from the Sun plus the distance of Klaatu's world from Earth when these two planets are at inferior conjunction (i.e., when they are aligned together on the same side of the Sun and as viewed from Klaatu's planet), we can arrive at an approximate distance (mi) away from the sun of

$$93000000mi + 250000000mi = 343000000mi. \tag{2.1}$$

This means that, with Klaatu's world having a semimajor axis of approximately three hundred forty three million miles (See Figure 2.1) and since Klaatu's home planet is just two hundred fifty million miles away from Earth (instead of being two hundred fifty million *light years* away in contrast, for example) Klaatu's world would have to be in orbit around our Sun and lying somewhere inside the asteroid belt. Then his home world would be about 3.688 astronomical units (AU) away from the Sun. That would put his home world unrealistically somewhere inside the asteroid belt just beyond Mars orbit and somewhere inside Jupiter's orbit at a distance in AUs from the Sun of close to 3.7. Why do we say *unrealistically*? Because Klaatu's planet would have to lie somewhere inside a Kirkwood gap between the orbits of Mars and Jupiter at about

$$\frac{343000000}{93000000} = 3.688, \tag{2.2}$$

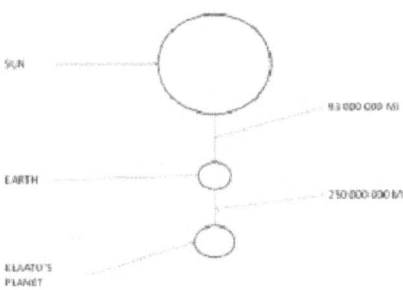

Figure 2.1: Klaatu's hypothetical planet in conjunction with Earth.

or about $3.688 \approx 3.7$ astronomical units away from our Sun, which means Klaatu's planet would have an orbital resonance near or very near 5:3. This means by Kepler's Third Law that Klaatu's planet would complete nearly five orbital periods around our Sun to three orbital periods completed by Jupiter.

Now this latter possibility, that is, that Klaatu's planet is about 3.7 AUs away from the Sun and lies in the asteroid belt possibly somewhere inside some Kirkwood gap with an orbital resonance of 5 : 3, is a contradiction, because there are no large planets known to exist either within the asteroid belt or within any of the asteroid belt's Kirkwood gaps. Why do we know there are no such large sized planets in the asteroid belt that would have such an orbital resonance? Because for one thing, if Klaatu's world was in the asteroid belt at a distance of nearly 3.7 AUs away from the Sun it would be inside or very near a 5 : 3 Kirkwood gap, which means it would have a planetary orbit that is unstable and chaotic. Also at present we know that Ceres is the largest asteroid in the asteroid belt, and this asteroid, which is about 2.768 AUs away from the Sun, lies within the semimajor axes distribution between orbital resonances 3 : 1 (at 2.5 AUs) and 5 : 2 (at 2.8 AUs) but not inside or near a gap, and at present there are no large planets or asteroids known to exist at 3.68 or roughly about 3.7 AUs away from our Sun and that have also stable orbits. What is more is that although Ceres, at about two hundred fifty seven million miles away from the Sun[7] is the largest asteroid known to exist in the asteroid belt it is much smaller in mass than is our Moon! So if Klaatu's planet was instead an asteroid as massive as Ceres or even smaller within the asteroid belt instead of a large planet like Earth or Mars, it would have to be a very low gravity planet or low gravity planetary dwarf, which suggests that Klaatu and Gort would have had very serious problems trying to adjust down here to Earth's much more intense gravitational field the minute they both stepped outside their flying saucer.

So even though Klaatu's planet could not be two hundred fifty million miles away from Earth orbiting the Sun at a distance of about 3.688 AUs from somewhere within the asteroid belt, we still can speculate however about the technology level of Klaatu's civilization if we pretend that his planet was much farther away from Earth, that is, in

---

[7]Do not forget that Klaatu claimed his world was two hundred fifty million miles away from *Earth*, not two hundred fifty million miles away from the Sun. Yet in our example Ceres is close to being two hundred fifty million miles away from the Sun.

terms of *light years* for distance, instead of distance from the Sun in miles or astronomical units.

So let us suppose instead that Klaatu's home world was two hundred fifty million light years away from Earth instead of being only two hundred fifty million miles away from the Earth. We know from the film that Gort can melt away any material in which he was encased by military scientists, including all known metals, metal alloys and plastics on Earth. This means the aliens would have to have a very deep understanding of solid state physics. Also the robot can destroy the Earth, as Klaatu claimed. What if Gort did have such power? With sufficient impact velocity an asteroid of sufficient mass and about sixteen kilometers (9.9378 miles) or more in length or width could annihilate all life and civilization on Earth as well as ignite a fire of enormous and global proportions. Such an asteroid impact event would unleash energy at least something close to 4.18×10^{23} Watt-seconds. So let us assume that Gort could unleash this much energy to destroy the Earth. Therefore if Klaatu's home world or planetary system had a minimum power level of the order of 10^8 megatons per second where this is about the amount of energy unleashed if a sixteen kilometer long asteroid where to hit Earth, we can use a formula provided by the astronomer Carl Sagan, to estimate where Klaatu's civilization might fit on the Kardeshev scale:

$$K = \frac{\log_{10} P - 6}{10}, \tag{2.3}$$

where P denotes power in Watts. Using $P \approx 10^{23} W$ at the very least, we get

$$K \approx 1.7. \tag{2.4}$$

This would mean that, in terms of science and technology levels, Klaatu's home planet would have to be, at the very least, somewhere between a Type I and Type II alien civilization on the Kardeshev scale if not higher. Klaatu's world would be a place where his people

1. Obtained control of their planet's weather and climate systems,

2. Had the capability of developing new metallic alloys that have astounding material properties,

3. Designed and built robots capable of piloting interstellar spacecraft, disintegrating human weapons like rifles and tanks into thin air (perhaps through applying some very advanced knowledge of particle physics) and destroying the entire planet Earth with a minimal energy equivalent to that released by the collision of a massive asteroid with Earth, where the asteroid is at least sixteen kilometers either in length or in width.

That is a lot of energy for one robot to have! It is inconceivable that any alien civilization would have been able to arrive at such an advanced technological level without first having eliminated anything that would have gotten in the way to prevent its civilization's advancement, or to postpone such high technological advancement for centuries more, including nuclear war, worldwide political, economic and religious discord, violent tribalism and antisocial or criminal behavior.

2.8 Tobor the Great

The astronomer Gerard Kuiper (1905–1973), whose mentor was none other than Ejnar Hertzsprung (a name well known to astronomers, astrophysicists and their undergraduate students in their physics courses as was I some years back, who are familiar with the meaning of "main sequence stars"), made significant discoveries on binary stars, the rings of Uranus and the presence of a methane atmosphere around Saturn's moon Titan. He also had considerable confidence in the ability of telescopes, unmanned space probes, autonomous weather satellites and remote controlled machines to collect information on objects in distant space, whether these objects are stars or planetary bodies in our solar system. Indeed based on how many of his discoveries depended upon various optical and radio telescopes, infrared sensors such as was used on NASA's Kuiper Airborne Observatory (KAO) back in the nineteen fifties and remote satellite space probe technology, one is left wondering if Kuiper had more confidence in insentient physical instruments and remote technology to gather astronomical information and data or in sentient human astronauts. This was the premise in the fifties film *Tobor the Great*, a robot related film I once saw as a kid but which I did not think was as good as *The Day the Earth Stood Still*.

In the 1954 movie *Tobor the Great* (Yes indeed, "Tobor" is just the mirror image of "Robot") there is a character named Doctor Ralph Harrison, a government scientist who is outraged at the excessive dangers to which prospective astronauts are subjected, such as exposure to high level g forces. Together Harrison and a colleague named Nordstrum design and build a robot named Tobor, the purpose of which is to spare human astronauts from the hazardous effects of space travel which would include among other things, high g forces, cosmic rays and collisions with small asteroids or meteors.

Tobor is extremely unique in that he is a semiautonomous machine, meaning in a sense that he is a "teleoperated" robot (We discuss more about telerobotics in a future chapter). He is capable of performing some physical tasks on his own, including walking up and down spiral staircases, wrecking furniture, clobbering communist espionage agents and typing. Nevertheless to complete other tasks he depends upon extra sensory perception (ESP) signals generated by the brain of a human. Here perhaps is where the suspension of disbelief might break down while the skepticism level might rise for some roboticists. Even so it must be said that human to robot or human to computer communication or interaction by means of brain or brainwave generated signals might not lie outside the realm of future technological possibility, given any crucial and relevant advances in computer science and in cognitive neuroscience.

2.9 *Forbidden Planet* and Robby the Robot

The movie *Forbidden Planet* introduced a robot that became something of a science fiction superstar in the nineteen fifties, sixties and seventies. In actuality Robby the robot was an elaborate root costume worn by an actor. Nevertheless the effect was so realistic in appearance the costume when worn and operated properly easily gave the impression it was a walking, talking mechanical man.

The setting is Altair 4, a planet where Doctor Edward Morbius and his daughter Altaira have survived a disaster that befell their fellow shipmates upon arrival to the planet when their starship was destroyed by some unknown alien power. Presumably Altair 4 orbits the star Altair, (or α-Aquilae, a spectral A type star on the Hertzsprung-Russell diagram). The plot of the film advanced ideas on Jungian and Freudian psychology, interstellar travel and robotics that were extremely imaginative, such as Jungian ideas on the collective unconscious, starships, mapping the brain (evidently the ancient race of Krells had done this in the film, not humans) and the neoplatonistic idea on *hypostasis*, such as when the character Doctor Morbius, played by actor Walter Pidgeon, creates unknowingly an actual violent monster from his own ID when he is sleeping. Ironically this latter feat in the movie brings a saying by the Spanish artist Goya to mind: *el sueño de la razon produce los monstruos*.

Based upon the network of nuclear reactors in the underground region of Altair 4, one could assume that the Krells were at least Type I on the Kardeshev scale. What is evident clearly in the movie is that Robby observes Isaac Asimov's Three Laws of Robotics. So when the Krell technology compels Morbius to threaten against his will not only the safety of the starship visitors from Earth but also the safety of his daughter Altair, he orders Robby to take action against the monster. Robby though is unable to execute this command, since doing so would harm Morbius.

2.10   *Lost in Space*

The TV show *Lost in Space* had a predecessor which was *Space Family Robinson*, a comic book published by Gold Key Comics. The plot focus of the TV drama was on a scientist and his family aboard a spaceship that gets lost in interstellar space after a meteor storm. John Robinson was the astrophysicist lost in space in the Alpha Centauri system aboard the *Jupiter Two* along with his biochemist wife June, their three children Will, Penny and Judy, The Robot, Space Corps Major and pilot Don West and the narcisstic, troublesome and conspiratorial character Doctor Zachary Smith. Will Robinson was the boy genius, his older sister Penny, sensitive and compassionate, both of them supervised on occasion by their patient older sister Judy. Cowardly, indolent and with his utterly stentorian and stilted manner of speech, Doctor Zachary Smith was the villainous and selfish saboteur in the show, who managed always to get the intrepid family of spacefarers into one serious dilemma after another one.

The first few seasons were spectacular and compelling, at least for a kid thirteen years old as was I when I watched the program every week. That season some of the problems they confronted seemed to exhibit sufficient suspension of disbelief: ion storms, novas, the threat of food shortage and computer and equipment failures on the ship.

One realistic part of the show was The Robot. It had a good command of natural language. It could walk, wheel itself about and talk. The Robot followed orders, warned whenever some danger was imminent and could make life saving decisions on the spur of the moment. It also was a good judge of human character, since it was capable of reaching through its own logic the conclusion that Doctor Smith was an insufferable cad.

The Robot in *Lost in Space* in some respects was not dissimilar to C3PO in the movie *Star Wars*. In fact it is conceivable that such robots, if they do not exist today, probably could be designed and constructed within the next few decades.

2.11   *Silent Running*

We mention this film by director Douglas Trumbull because the plot also focuses on the types of robots that humans will be building more than likely within the next twenty or thirty years if not sooner. In fact already there are service type robots in production performing various roles, such as security surveillance within buildings, the handling of dangerous explosives and light janitorial services. *Silent Running* is a bleak cautionary tale about the triumph of human greed and commercialization over the survival of plants and how one human botanist along with three small bipedal service robots, strive to preserve a small portion of Earth's multifarious plant organisms on a freighter near the Jovan planets.

When space traveler and Botanist Freeman Lowell is ordered to destroy all the plant life in his biodomes on the *Valley Forge* and return the freighter to commercial deployment he rebels by seizing control of the freighter and proceeding deeper into space, in a mission to preserve the plant life in the last remaining biodome. After various activities and accidents, two of the service robots are lost. The last bipedal robot remaining named Dewey becomes the caretaker of all the plants in the last remaining biodome, fulfilling its task daily for as long as is possible.

Already there exist robots in operation that can vacuum rugs, remind elderly patients or hospital patients to take their medication when the time arises to do so or that can play ping pong with a human opponent. Asimo is an example of a service robot that can be something of a receptionist as well as an automated guide and diplomat. A small bipedal robot like Dewey, an autonomous service robot that can tend faithfully to routine gardening activities and plant cultivation does not lie outside the realm of possibility. One can imagine it could run various machine learning algorithms in order to manage successfully many kinds of unforeseen contingencies, like severe weather conditions, accidental damage due to human activity or plant pest infestation.

2.12 Intelligent Machines from *Star Trek*

2.12.1 Androids and Harry Mudd, Captain Kirk's amoral, interstellar Space Gadfly

In the *Star Trek* episode "I, Mudd," Harcourt Fenton Mudd, who in the previous episode "Mudd's Women" proved himself to be an antisocial pest and gadfly in his run-ins with USS Enterprise Captain James T. Kirk, is a space criminal and fugitive who escapes capital punishment on some alien world by hijacking a spaceship to escape to another planet populated by hundreds of thousands of gorgeous, anatomically female resembling androids. However amid all these thousands of machines there only is one male android named Norman who is some sort of master coordinator for all the intelligent machines. Norman uses deception by passing himself off as an Enterprise officer to commandeer the ship and to take it and its human crew to his planet to which Mudd had escaped. Once Kirk and the others arrive they are told by Norman that the human proclivity for violence makes humans dangerous. Therefore the androids plan to enforce all the crew members to abandon their ship to live on the planet so that the androids can "take care" of them. Kirk foils Norman's plans by getting the captured human crew members to make a sequence of statements and physical actions that contradict each other. For instance the starship crew members begin to dance although the androids hear no music. Ship engineer "Scotty" feigns sickness and "dies" before the very bewildered android Norman as Scotty suddenly proclaims he died from "Too much happiness." Then Captain Kirk delivers the *coup de grace* when he tells Norman "Everything I tell you is a lie. I am lying."

When it comes to intelligent machines concluding on their own that they must control humans because humans are too destructive, this particular *Star Trek* plot is reminiscent of the plot in the nineteen sixties novel *Colossus: The Forbin Project*. In that novel an artificially intelligent American supercomputer named Colossus has control of all the country's thermonuclear weapons systems. After it links with a similar Soviet machine named Guardian the resulting composite supercomputer uses its stewardship over all nuclear weapons in America and Russia to dominate the world, having concluded humans cannot be trusted to run the world.

Can this sort of thing really happen, depending on the level of artificial intelligence given to future machines? In order to enforce any paradigm of US versus Them, would such artificially intelligent machines have to have reached some level of consciousness and self-awareness? If the answer is yes then preventing certain intelligent machines from reaching such a level of artificial intelligence in the first place might be one way to prevent the disaster of mechanized domination. Another option might be to build some type of fail safe, back door application in any future AI machines that have the capacity to self-evolve their level of artificial intelligence.

Another thing to consider with regard to the android Norman and his cohorts is that as we mentioned, Norman is some kind of "master coordinator." When his systems malfunction all the other android systems malfunction. It is not inconceivable for a future, highly intelligent and swarm robot system to behave this way. If any one robot "stampedes" so may all of them follow the same example. Even distributed networks of machines that are not intelligent will malfunction if one or several specific machines in a network malfunction. Take for example a network of master and slave DNS servers. If the master DNS server machines all malfunction at the same time, such as during a power outage or a severe weather event, then the DNS records in the slave machines cannot be updated.

Another example is a scale free network of servers. If certain machines are "knocked out" along the scale free network then a large portion of the network can be crippled.

2.12.2   The Ultimate Computer

One memorable episode from the original television series *Star Trek* was the episode called "The Ultimate Computer." In one sense it was memorable because it served as a warning to those who see no danger in building machines that think like humans think.

Scientist Richard Daystrom builds the fifth version of a multitronic computer called the M5[8]. The first four versions failed to operate properly. The new intelligent computer is installed into the Enterprise to take over the starship's command and control systems. The new computer system is so efficient it becomes apparent to Captain Kirk (called "nonessential personnel" by the M5) that he might be forced into an early retirement while the new computer assumes all the duties and responsibilities of the starship captain.

Daystrom designed and built the M5 so that it operated as does an intelligent human, with both a reasoning and decision making capability that allowed the computer

---

[8]The expression "M5" happens to stand also for a particular kind of mathematical message authentication algorithm (MAC) used in computer and network security purposes. However this was long after the original Star Trek episode was aired, when it used the designation M5 to refer to an intelligent computer rather than to a message hash or message authentication code.

to perform intricate tasks, to win in various military tactical and battle game situations and to make complex decisions swiftly. The machine's memory core was built around the concept of human memory engrams, something that reminds this author of the reinforcement learning in some artificial neural networks that are so useful today in artificial intelligence research.

However as human memory can be shaped by human experiences along with various human emotions, the M5 Multitronic Unit begins to display bizarre behavior, indicating subtle errors in its programming. First it shuts down life support systems at various locations on the Enterprise everywhere no human crews are working. It utilizes power not from its own power supply but directly from the ship's warp drive. Then when the Enterprise must participate in war games sponsored by Star Fleet, the M5 uses its command of the Enterprise weapons systems, to destroy the starship *Excaliber* and the ship's entire crew that was participating in the war games. Later a human crew member is disintegrated by the M5 when Kirk gives the order to 'pull the computer's plug.'

When Daystrom becomes mentally unstable after continued questioning from Spock, McCoy and Kirk his invention also runs berserk, while Daystrom's need for survival is transformed into his invention's need for the same thing.

If there is a motto to this story it is "Garbage in, garbage out." If one day we do build robots and intelligent machines that emulate or mimic human emotions and reasoning abilities, then we should not be surprised if our own intelligent machine creations exhibit in some way our more distasteful human traits, like human prejudice, egotism and even violence and recklessness enlisted in the cause of self preservation like Daystrom's multritronic M5 computer.

Gene Roddenberry might have had one marvelous vision about humanity's future destiny in interstellar space with intelligent computers serving aboard "warp drive" starships. However the present sorry state of human affairs in America and the world makes all that future highly dubious. It is extremely unlikely that humans even will succeed in launching von Neumann probes out beyond the confines of our solar system to trek "where no one has gone before." For *H. sapiens* it would appear that national and international conflicts are far more likely to be events in the future than any serious international preparation for space travel and eventual colonization.

2.12.3 Gene Roddenberry's Data, the Android who would be Man

In the science fiction TV series *Star Trek: The Next Generation*, which aired in the nineteen eighties and nineties, the android "Data" is described as being "positronic," and in the episode "Phantasms" there is a scene between the android character Data and the human starship engineer named Geordi LaForge in which the android admits he processes algorithms by means of a "neural net," something we will consider in Chapter 11. One only can wonder if his neural net is some technological advancement over deep learning or reinforcement learning. Data had artificial intelligence that by far superseded the biologically based intelligence of humans. He was swifter, incredibly dexterous and nimble and could simulate the voice of any other human should the need arise. Yet ironically Data admired humans and yearned to be human. The term "positronic" first was popularized by science fiction author Isaac Asimov, who got his start as a prominent science fiction writer under the tutelage of a fascinating, highly opinionated, controversial and innovative SF editor and author named John Campbell. As we have seen already in this Chapter, Isaac Asimov explored the many possibilities of intelligent robots in several science fiction stories and articles. He has become the recognized inventor of what he called the Three Laws of Robotics, a sort of protocol that governs and restricts robot behavior when these machines interact with humans.

2.13   HAL 9000

Today probably much more is known and remembered about HAL[9]. than about many of the other artificially intelligent machines we have discussed so far in this chapter: An artificially intelligent computer that could speak and understand human speech, read lips, play chess and mainly function overall as an intelligent command and control computer system aboard a human spaceship. Anyway HAL runs berserk, killing crew members and then fighting Dave, the lone intrepid astronaut who tries to uninstall the rogue computer as the computer sings "Daisy Bell."

2.14   A Robot Grandmother and a Maid

One episode from the original television series of *The Twilight Zone* presented us with a script based upon a short story by science fiction author Ray Bradbury. The television script "I sing the Body Electric," had a title that was based upon the poem by Walt Whitman. The episode depicted a grieving father and his three small children as they trekked together to a store named Facsimile Limited. This store supplied customers with android robots of various types and the father had made arrangements to buy a matronly, friendly female android named Mrs Hutchinson. The android was built to resemble the loving, elderly grandmother who had died recently.

Caring and endearing, the female android wins the hearts of two of the children. The third however, a daughter who is the oldest of the three kids, remains distraught and depressed over the death of her human grandmother and rebels, refusing to accept the mechanical facsimile who replaced the real grandmother. This girl undergoes a change of heart however and she accepts the android at last when the matronly female android saves the girl's life in an automobile accident. The android "grandmother" remains an intimate part of the family until the three children grow old enough to leave home for college.

But with the episode "You can't get Help like that Anymore" from Rod Serling's television series *Night Gallery,* the viewer observes the unraveling of an entirely different sort of plot. Two characters, Mister and Mrs Fulton, have purchased an android female maid from a company that manufactures domestic robots. They return to the plant with the maid an absolute physical wreck, as the couple complains to the manager about the company's 'lousy' products. The manager Mister Hample notices that the android maid was the victim of vandalism of the most vicious sort and that the Fultons are responsible for the destruction of the robot maid. Furthermore the two company employees, Mister Hample and the company roboticist Doctor Kessler can see that the female robot maid had cried before she was vandalized, since she displays tears on her face.

Mister Hample provides the Fultons with another maid. But then the couple return home to reveal their true natures to the new robot maid, who is just as attractive as

---

[9]From the novel and subsequent film *2001: A Space Odyssey,* based upon the novel by science fiction author Arthur Clarke

was the first one before it was destroyed. In truth Mister Fulton is an abusive lecher and a drunk who subjects the android female maid to sexual harassment and ridicule. Mrs Fulton is as vicious as a crone. She subjects the new maid to vicious verbal abuse and threatens the machine with physical injury. Indeed Mrs Fulton does turn violent against the android maid, but the two humans are shocked when the machine fights back successfully to defend itself. The episode ends with the androids taking over the company.

These two different episodes are a study in contrasts. The first episode from *The Twilight Zone* showed how a robot can learn good behavior from humans, since the Mrs Hutchinson android had learned well from the endearing grandmother's personality. On the other hand the episode from *Night Gallery* displayed a very likely future scenario of an intelligent robot using violence in self defense against two human aggressors from which it had learned only too well about the more sinister and brutal side of human nature.

Some roboticists claim that in the next fifty years it will not be impossible for intelligent robots to emulate human emotions. We will consider the possible impacts of this in future chapters, including the possibility of artificially intelligent malware being a reality in the future.

2.15 *Blade Runner*

Blade Runner, a film based in part on the 1968 Philip K. Dick novel *Do Androids Dream of Electric Sheep?* and directed by Ridley Scott, is a bleak and pessimistic movie about a future planet wide human dystopia in which scientists have created a race of genetically engineered androids to perform hazardous work in offworld colonies, where most of the planet's animal species have become extinct and where multinational corporations have the power of nations. Within this morally anemic world of the future there is a class hierarchy between the humans and the genetically engineered replicants who, with lifespans that are programmed deliberately to be excessively short, are forced to serve them. There is a tragic irony in this, since it is the inhuman replicants themselves who value ideals such as freedom and life while their human masters have become dehumanized. The film also raises a troubling question: Is this sort of thing likely to happen in a future world in which humans and artificial life forms or robots interact? That should not be a question too hard to answer, given the last ten thousand years of human history. The novelist William Faulkner once claimed history is not *was*, but *is*.

One thing that almost all science fiction robots seem to have as they interact with human characters inside countless novels and short stories of science fiction as well as on the movie screen is intelligence. But how does one define "intelligence" when it comes to machines? Are all computers, for example, intelligent? This is important to understand within the context of computer science and information technology especially if someone uses the word "intelligence" to describe certain types of databases, virtual humans or automated space probes. After all a scientific calculator can perform arithmetic operations, or it can be used to plot a histogram, a parabola or a limaçon using polar coordinates, to compute the secant of $27° 16' 48''$ in radians, or to find the integer remainder left when one divides 33^{16} by 17. Yet these computational abilities do not make a handheld calculator "intelligent." Both an abacus and a handheld calculator can sum a list of numbers, but neither of these devices can explain to you in natural human language and in a reasonable and sensible manner why a Beethoven piano sonata is better than a candy wrapper.

On the other hand there have been laboratory rats trained successfully to navigate through a maze and gorillas and chimpanzees who have been trained to communicate with human researchers through a sophisticated means of sign language. But then chimpanzees and gorillas are more intelligent than calculators. Keep in mind a calculator cannot communicate with biologists, or navigate its way successfully through a maze. But a computer that could understand English text or speech and hold down a conversation with a human would be a computer that does exhibit intelligence. So in the next chapter we will consider how computer scientists began to study how computers could understand language by creating for the abstract computer something called a formal language.

2.16   K. I. T. T.

Can an automobile be intelligent?

Before there were self-driving cars there was the self-driving car K.I.T.T. on the eighties television show *Knight Rider*. In the series the character Michael Long, a police detective shot in the face while on duty, is subjected to plastic surgery then reappears anew as the mysterious Michael Knight, an undercover operative who works for the Foundation for Law and Government, a secretive organization financed by a billionaire named Wilton Knight.

Operative Michael Knight, working under the orders of his supervisor named Devon Miles, is given a new partner named KITT, an acronym for Knight Industries Two Thousand. KITT is a Pontiac Trans AM like no automobile on Earth. Designed by brilliant mechanical engineer Doctor Bonnie Barstow, KITT runs and operates by a computer endowed with a very sophisticated artificial intelligence program that uses natural language understanding, allowing Knight to get KITT to respond immediately to a variety of spoken commands, to gather and to collect crucial information and even to exchange wisecracks.

But the automobile goes even beyond all this. It can run algorithms to enable it to take its own initiative whenever Knight is in serious trouble, such as if he is captured, tied up or about to be killed by criminals or enemy agents. If Michael Knight escapes from a building in a hurry then KITT knows how to show up in time outside a nearby door or window. If the human operative needs to get the police KITT knows how to call them in a hurry.

Today we already have seen the appearance of self-driving cars. It is very possible we could see automobiles that can do things similar to what KITT could do within the next few decades.

Those who would like to know more about the TV show along with fans, can learn more at

`http://www.nbc.com/knight-rider-classic?nbc=1`

.

So just what is "intelligence," especially when this term is applied to machines that can "understand" human speech? We consider this question in subsequent chapters.

Chapter 3

Turing Machines

3.1 Who is Who: Will the real Alice *please* stand up?

It is Friday morning at nine o'clock within some US business office of some high rise office building, where we have some conversation as follows:

Bob: Alice, did you confirm my plane reservation for
                       this afternoon?

Alice: Sure did. Flight 137 leaves for London at two PM.

Bob: Super! I knew I could depend on you.

Alice: Of course. Do I ever let you down?

Bob: No, sweetie.

Alice: Hey, better watch that. You're married.

Bob   (Laughs):   I know. Just teasing.

Alice (Indignant):   Harrumph. I'm sure Maryann would not appreciate
                       you flirting with me like that ....
                       I am quite attractive you know!

Bob: Better call my wife after I leave for the airport
                       this afternoon.

Alice: So .... I estimate that will be around one?

Bob: Yep.

Alice: OK, I'll call Maryann at ten minutes after one
                       then.

```
Bob:              Great.

Alice:            Who else should I notify?

Bob:              Call our client. Tell him I'll meet him at
                  Heathrow when I get to London.

Alice:            Will do, boss!
```

Can a computer sound like Alice? This type of "natural language" conversation between a corporate executive and his administrative secretary inside some city based high rise office building might not be all that uncommon. But what if Bob was a human being while "Alice" was really a computer?

Many of us today are familiar with telephone based speech recognition systems. "Your call is important to us," we hear as we wait to obtain some crucial information by telephone. "Please stay on the line and your question will be answered by the next available operator." Then there is: "If we have answered your question please say yes and hang up. If we have not answered your question, please say no or press one," etc.

But what really is happening here? We might say YES or NO, whatever the case might be, in response to the questions or instructions given by the machine at the other end of the phone line. Yet what it really is doing is recognizing various spoken words that follow, more or less, predictable speech patterns. More than likely if you were to say the following into such a telephone based speech recognition system, "Hey! Just tell me why my phone bill is so high this month, will you?" You might get a response like the following one:

```
I'M SORRY. I AM HAVING TROUBLE UNDERSTANDING YOU.
```

But Alice, clearly if it is a computer secretary or embodied conversational agent in contrast to a human agent, had no difficulty apparently, in understanding Bob, despite any natural language patterns or idiosyncrasies that Bob might have revealed himself to have in their conversation. In fact if Alice in the conversation really is a computer, its responses did not differ at all from most predictable verbal responses that might have been given by a real human secretary. In fact if some other human being had standing outside the office door to listen to the conversation between Bob and Alice on the other side, he or she might conclude that it simply was a conversation between two human beings, a human executive and his human secretary.

Yet does the mere fact that Alice seemed to make responses that reflected a deep understanding of what her boss was saying with natural speech, mean that Alice *understood* every word that Bob spoke?

The answer is no, not necessarily. In fact Alice might have been a computer without even the intelligence level of a cockroach. Cockroaches of course, do not understand natural human languages. They are not intelligent enough to check plane reservations for corporate employers, or to chide that employer for his flirtation.

AIML, which stands for the *A*rtificial *I*ntelligence *M*arkup *L*anguage, is a programming markup language that is a subset of sorts, of XML. As he discussed in his book, *Virtual Humans*, Peter Plantec mentions how with AIML a software or web developer can design "embodied conversational agents," virtual humans capable of holding conversations with real humans with such seeming remarkable aplomb it appears as if the virtual entity actually has intelligence. This however only is an illusion. What AIML really does is to help one to design a kind of impulse and response system, where the virtual human agent makes spoken responses to verbally spoken inputs in a way that simulates a conversation between two people remarkably well.

What does it take then, to get a computer, virtual human agent or robot actually to understand a human spoken natural language when it "hears" it spoken? Already some computers have computer vision performing so well ingrained within their systems hardware and software they can help to identify the faces of terrorists in a crowd. Can a computer do something similar with human speech?

3.2 Alphabets, Vocabulary, Grammar and Language

A kitchen toaster cannot understand a human being's spoken sentence, something that has high levels of syntactic and semantic complexity. Neither can an iPad, a remotely controlled garage door or most building HVAC systems, thermostats, computers and robots understand human language, for that matter. In fact there are thousands of natural languages past and present, some of which are so complex such as Urdu or Navajo, that linguists have spent most of the time in their careers studying them. More than one half century ago linguist Noam Chomsky at MIT sought some kind of way to model the various syntactical properties of certain languages, with grammars that are among what computer scientists call *phrase structure grammars*. Chomsky and other linguists back then made an effort to get a much better understanding of many different kinds of languages, such as from English, German and Mandarin to Algol, COBOL and FORTRAN. Chomsky formulated many of his very insightful ideas on this topic in two important papers, "On Certain Formal Properties of Grammars," and "Three Models for the Description of Language."

So...what is exactly, a "phrase structure grammar," one asks?

First let us establish what theoretical computer scientists mean when they use the term "language" and in particular formal language. When they speak of "language" they do not mean exactly languages that include both the syntax and semantics for any human language from Armenian to Xhosa. Rather they use the word "language" in an abstract sense, in that these abstract or formal languages all fall into certain categories determined by their grammars, and these more abstract languages have clearly defined descriptions.

Thus computer scientists might talk about a language that has a *regular grammar*, or a language with a *context sensitive grammar*, or still another one that has a *context free grammar*.

In computer science a "language" (denote it by the symbol L) really is a vocabulary subset, that is, a subset of words taken from a larger set of all possible words that can be constructed from the alphabet. Let us clarify what this means because sometimes the words might not make sense in human languages, since if the alphabet is for example

$$\{a, b\}, \tag{3.1}$$

Then one of the legitimate words from L possibly could be something like

$$aaaaaaaaaaabbbbbbbbbb \in L, \tag{3.2}$$

or
$$abababababbababababaab \in L. \tag{3.3}$$

On the other hand if the alphabet instead is $\{0, 1\}$ then we can have words in L that are a binary string like
$$000111000111000111000111 \in L. \tag{3.4}$$

Now one can build the much larger set of all possible words from the language L, by a particular operation called *concatenation*. This means simply that if you have two different strings from L, like
$$con, \ catenate, \tag{3.5}$$

from L, you can combine these two "words" to make the larger word *concatenate*. Java programmers for instance do something similar to this all the time.

Suppose now that
$$L = \{ab, abb3, ax3c\}, \tag{3.6}$$

for some alphabet $\{a, b, c, 3, x\}$. Stephen Kleene, a mathematician who along with Alan Turing was one of the students of Alonzo Church, came up with a means to build the larger set of words from the smaller set L, by using the property of concatenation of words to define something today called the "Kleene closure" (some readers used to set theory might recognize some of this notation. Others please bear with it for the moment.)

$$L^* = \bigcup_{j=0}^{\infty} L^j. \tag{3.7}$$

To illustrate with the language L,

$$
\begin{aligned}
L^0 &= \{\epsilon\}, &&\text{(3.8)} \\
L^1 &= \{\epsilon, ab, abb3, ax3c\}, &&\text{(3.9)} \\
L^2 &= \{\epsilon, abab, ababb3, abax3c, abb3ab, abb3abb3, abb3ax3c, ax3cab, ax3cabb3, ax3cax3c\}, \\
L^3 &= L^2 L^1, \\
L^4 &= L^3 L^1, \\
\cdots &= \cdots &&\text{(3.10)} \\
L^j &= L^{j-1} L^1, \cdots &&\text{(3.11)}
\end{aligned}
$$

$j = 1, 2, 3, \cdots$ etc. by concatenation, where "ϵ" denotes a "null string," which in a programming language like C could be expressed by two adjacent beginning and ending quote marks with nothing between them, like "".

But languages also need their grammars. Many also need rules of syntax. In computer science a "phrase structure grammar" has five things on which to base a language's properties:

1. A set V of words for the vocabulary,

2. A subset T of V for *terminals*,

3. A subset N from V for *nonterminals*, where $T \cap N = \emptyset$,

4. A *start* symbol S,

5. A set P of *productions*.

A "production" is a set of grammar rules of syntax that can appear as ordered pairs of terminals and nonterminals, in one case. To illustrate,

$$\begin{aligned} sentence &\Longrightarrow & noun \quad verb, \\ noun &\Longrightarrow & ELVIS \\ verb &\Longrightarrow & ROCKS \\ punctuation &\Longrightarrow & ! \end{aligned}$$

where the nonterminals are "sentence," "noun," "verb," "punctuation," and the terminals appear in the string "ELVIS ROCKS!"

Productions are vital. With them one can arrange nonterminals and terminals into tree shaped data structures, where the leaves at the bottom are the terminals, to comprise something that can represent a parsed human sentence, depending upon the syntax rules for the grammar in question.

Before we saw that there are different kinds of grammars, like regular grammars, context sensitive grammars and context free grammars. A regular grammar has something called regular expressions, a term which doubtless many systems administrators on Unix or Linux server systems who write Perl or "shell scripts," might be well familiar. For instance

```
grep 'claude.*nnon' oldfile > newfile | tr 'claude' 'Claude'
```

tells the Unix shell to look for all lines in the file *oldfile* that contain the string "claude" followed by zero characters or any other number of characters up to and including "nnon," store the output into a file *newfile*, then pipe this to the shell command **tr** to change the string "claude" to the string "Claude" in *newfile*.

A "regular expression" is a way to "build up" a regular language, by using an operation to generate new strings or words. For instance,

$$L(a^*) = \{\epsilon, a, aa, aaa, aaaa, aaaaa, \ldots\}. \tag{3.12}$$

Since many computer programming languages exploit the properties of certain phrase structure grammars, compilers and interpreters have been built for them. These compilers and interpreters help to parse written computer programs into different subsets of strings, before the code is converted into machine code that the computer can understand. A deeper understanding of context free grammars and context free languages also can enable artificial intelligence and natural language processing researchers to design future software systems that not only will lead to better computer language translation endeavors, but also to the design of computer and robot systems that will understand human language far better than do most of such systems currently in use.

All these different kinds of grammars make up a sort of hierarchy. A regular grammar is a proper subset of the set of context free grammars, which in turn is a proper subset of context sensitive grammars. The context sensitive grammars are a proper subset of recursively enumerable grammars. This latter grammar is relevant to that abstract computer that has come to be known over time as the Turing Machine.

3.3   Abstract Automatons and Turing Machines

In 1956 Stephen Kleene proved a theorem that turned out to be very useful in the context of abstract machines that model the behavior of a computer or data processor. He proved that a finite state automaton can recognize a regular set. Now since these regular sets are generated from regular grammars, the theorem more or less means that a finite state machine can recognize if a given phrase structure grammar for some language is regular or not. In fact there are also other abstract machines like the Turing Machine that can recognize any recursively enumerable language.

To give the precise and formal mathematical definitions for "finite state automaton" and "finite state machine" is beyond the scope of this book. Suffice it to say that both a finite state automaton and a finite state machine are two abstract machines that can input words of a language to help it to move from one state to another state in a finite succession of state transitions depending upon the length of the string, in order to perform some task that leads the automaton to some kind of final state. However there is a subtle difference in how these two machines perform their tasks, because a finite state machine can do some things that a finite state automaton does not do, like generate output strings from input strings. By "language" we do not mean necessarily French, Russian, Korean or even C, Java or Python. We mean instead an abstract kind of language. But what we have described here is for what is called a "deterministic" finite state automaton or machine. There also is another type of finite state machine called a "nondeterministic" finite state automaton or machine. This other kind of automaton or machine has the *power to make decisions* at each step, in the sense that it is not restricted to move only from one single state to another state. At each transition or step in the process it can "decide" upon what the next state should be to which it shall move. An analogy for this would be a chessboard. Imagine that all the red and black squares are all numbered consecutively. If it finds itself on some red or black square having some specific number, then depending upon the specific square it is on it could have the option of moving to any one of two or three other squares of the same color.

Finite state automatons and finite state machines are everywhere today in civilization. You might be surprised to know that many of the most simply constructed actual physical machines today, light switches, vending machines, ATMs, Internet routers, switching circuits, firewalls, intrusion prevention systems and even computer programs, compilers and interpreters included, behave like the more abstract finite state machines or automata that can be described as "recognizing" a language. All one has to do is to remember that here we are using terms like "alphabet," "word," "grammar," etc., in a very abstract sense. So for instance if a candy vending machine in the cafeteria of a high rise office building accepts nickels, dimes and quarters for a total of sixty five cents for delivering a candy bar, we can denote these by "N" for "nickel," "D" for "dime," "Q" for "quarter," to get various words that the finite state automaton will recognize, like

$$NNNNDDDD, \ NNNNNNNNNNNN, \ QQDN, \ DDDDDDN, \ldots \quad (3.13)$$

and we can come up with productions and with *state diagrams* to illustrate the behavior of the particular finite state automaton. Today software engineers even utilize diagrams of finite state machines from the Uniform Modeling Language, to model different aspects of software systems or architectures they might be designing.

Whenever we use a candy vending machine, an electronic door that opens when we stand on a footpad or even a token, coin or card operated turnstile at a transit station, we are interacting with either one of these two types of machines.

Let us focus for a moment on the finite state automaton, which we shall abbreviate here as FSA. We can for instance have two states in the FSA, namely START and ACCEPT. Let START be what we call the initial state and ACCEPT the final state. Five things are needed for us to have an FSA, a set for the two states for our example,

$$S = \{START, ACCEPT\}, \tag{3.14}$$

an initial state which here in our example simply is START, a subset for the final state which we can denote as $Y = \{ACCEPT\}$, a set A for our alphabet to get the "words" of a formal language, which here shall be

$$A = \{0, 1\}. \tag{3.15}$$

Finally we need something called a *transition function* F, which has a domain of ordered pairs from the two sets S and A and a codomain or range in the set S. Thus we have a rule of assignment

$$F : S \times A \to S. \tag{3.16}$$

We have also our abstract FSA then, denoted by the set

$$\{A, S, Y, START, F\}. \tag{3.17}$$

The Cartesian product set $S \times A$ really is a relation with ordered pairs

$$\{(START, 0), (START, 1), (ACCEPT, 0), (ACCEPT, 1)\}. \tag{3.18}$$

We have even a state transition table for our FSA and a directed graph which displays the process.

STATES	0	1
START	START	ACCEPT
ACCEPT	ACCEPT	ACCEPT

State Transition Table.

What are some of the formal language words, or strings, that can be accepted by our particular FSA? There are infinitely many possible words. Here are just some of the strings (See Figure):

$$\begin{array}{lll} 1 & 10, & \tag{3.19} \\ 01, & 100, & \\ 001, & 1000, & \\ 0001, & 10000, & \tag{3.20} \\ 00001, & 011, & \tag{3.21} \\ 010, & 11001, & \\ 010001, & 00011110001111110, & \\ 000\cdots, & 000000\cdots111111\cdots & \tag{3.22} \\ 01, & 011111111\cdots & \tag{3.23} \end{array}$$

These abstract machines as we have seen already go by different names, such as deterministic finite state automata and nondeterministic finite state automata that can recognize regular languages, and "pushdown automata" to recognize context free languages. But one of the most sophisticated types of these abstract machines is the Turing Machine, which was an abstract model of a computer which Alan Turing designed, to help him to answer the question as to whether or not computers would be able to think. A Turing Machine is able to recognize a special kind of language called a "recursive" language.

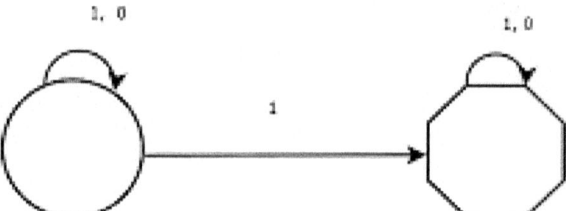

Figure 3.1: FSA for the state transition table.

But as we learned in the previous chapter where we encountered L. Frank Baum's character Tik-Tok, it is clear that Turing was not the first human being to wonder if machines could think. Baum himself was not even the first to ask this question. Even before Turing designed his abstract computer and Baum penned his stories for children, Ada Byron the Countess of Lovelace in her day pondered the question as to whether or not the Analytical Engine on which the mathematician and logician Charles Babbage had been working, could think.

3.3.1   Ada Lovelace, Bernoulli Numbers, Programming and "Astuteness"

Ada Byron (See the Wiki link for Ada Lovelace in the Preface), Countess of Lovelace (1815-1852), as the daughter of the English Lord Byron the brilliant but notorious poet, had developed at an early age, an aptitude for mathematics. These studies were encouraged by her mother, who had rebelled against her former husband's lascivious lifestyle. One of Lovelace's tutors was the mathematician August DeMorgan, a name well known to students who have delved into the theory of sets to learn about DeMorgan's Laws, in math departments either in high school or in college. One thing that is known well about Lovelace, at least in computer science circles, is that she designed a punch card based computer program for Babbage's Engine that included the use of Bernoulli numbers. Bernoulli numbers are certain rational numbers that crop up from time to time in various areas of number theory, physics and even in astronomy. They figure in the theory of ideals and in p-adic fields, which are branches of number theory. For instance mathematician and number theorist Ernst Kummer was able to prove that Fermat's Last Theorem

$$x^p + y^p + z^p = 0, p > 2, \tag{3.24}$$

where p is a prime number, is true provided p does not divide the numerator of a Bernoulli number. Such primes are called *regular* primes. In his proof Kummer looked at solutions to Fermat's equation over "cyclotomic fields," which are number fields (usually denoted in the algebraic number theory literature as $\mathbb{Q}(\zeta_p)$) that contain the zeros

$$\zeta_p \in \{1, e^{\frac{2\pi i}{p}}, e^{\frac{4\pi i}{p}}, \ldots, e^{\frac{2(p-1)\pi i}{p}}\} \subset \mathbb{C},$$

(where $i = \sqrt{-1}$) of the polynomial

$$z^p - 1 = 0, \tag{3.25}$$

where z is a complex variable.

Bernoulli numbers also crop up in atomic physics. In particular they are related to what is called the "Balmer series," which is a series of real numbers that was important in the early Bohr quantum model for hydrogen that theoretical physicist Niels Bohr had developed in the first twenty to thirty years of the twentieth century. The Balmer series determines certain emission spectral lines of atomic hydrogen that lie within the ultraviolet range. Emission spectra and absorption spectra help astronomers to determine what elements or molecules are present within a distant star or gas nebula.

Lovelace also was astute in having grasped a true understanding of the abilities of the Babbage Engine. In her notes about the Engine, which appeared in her translation of an article written by Italian engineer L. F. Manabrea about the Babbage Engine, she admitted:

> Considered under the most general point of view, the essential object of the machine being to calculate, according to the laws dictated to it, the values of numerical coefficients which it is then to distribute appropriately on the columns which represent the variables, it follows that the interpretation of formulae and of results is beyond its province, unless indeed this very interpretation be itself susceptible of expression by means of the symbols which the machine employs. Thus, although it is not itself the being that reflects, it may yet be considered as the being which executes the conceptions of intelligence. The cards receive the impress of these conceptions, and transmit to the various trains of mechanism composing the engine the orders necessary for their action. When once the engine shall have been constructed, the difficulty will be reduced to the making out of the cards; but as these are merely the translation of algebraical formul, it will, by means of some simple notations, be easy to consign the execution of them to a workman. Thus the whole intellectual labour will be limited to the preparation of the formulae, which must be adapted for calculation by the engine.

She understood well that the essential purpose of the Babbage Engine was primarily to calculate, not to interpret the data as does a human or a modern pattern analysis, knowledge discovery or machine learning system. Remarkably though there still have been some detractors, that is, scientists and other academic researchers as late as the nineteen nineties, who doubted Lovelace was capable of having a fully competent understanding of higher mathematics, despite the fact that not only was August DeMorgan her mathematics tutor who had expressed no serious complaints apparently about her mathematical abilities, but Babbage himself regarded her astuteness in mathematics as being innate and consequential.

In comparison to the prejudice Lovelace had to confront, it is doubtful Marie Curie even would have had her most important work published, if her own husband Pierre Curie had not insisted to the Nobel Committee in 1903, that the important results they had derived together largely was based upon her own work. At first the Committee had planned to ignore her results and to attribute the important results to both Pierre Curie and Henri Becquerel, who also had done research on uranium at the time. However Pierre and the Swedish mathematician Gösta Mittag-Leffler[1] fought for Marie to get the proper positive name recognition.

Human prejudices and biases die hard even among the highly educated, and sometimes they do not die at all. Human prejudice or bigotry on the basis of gender, religion, race, etc., is hard if not impossible to remove from one's disposition. Prejudice blinds any true believer, since the person who has it has the comforting assurance that he or she is unable to see it.

3.3.2   Alonzo Church, Alan Turing and Kurt Gödel

While Countess Lovelace clearly understood that the Babbage Engine was not a thinking machine, a "thinking" machine indeed was possible, provided such a machine could satisfy certain conditions, as Turing discussed in a 1950 paper called, "Computing Machinery and Intelligence."

The Turing Machine is an extraordinary contraption designed as an abstract structure. It is able actually, to "compute" in a similar way as does a human using pencil and paper. This is analogous to the computation of something with algorithms. The thing that encapsulates this idea is called the *Church-Turing Thesis*, which says in essence that a certain kind of Turing machine can do what a human does to compute with pencil and paper or what an algorithm does that is run by computer[2]. Alonzo Church developed what today is called the lambda calculus, which illustrates the behavior of certain kinds of computable functions that support the ideas of the thesis. Turing and Church sought a means to devise an "effective procedure" by which a Turing Machine could run an algorithm to solve a problem and to recognize a language, all in polynomial time. This means of course that the qualifying Turing machine can be identified with this would have to halt. In Chapter 5 of his book "Computation: Finite and Infinite Machines," Marvin Minsky at MIT provided a partial definition of "effective procedure" (I only say "partial" here because the actual definition is not so easy to state in just a few words):

> An effective procedure is a set of rules which tell us, from moment to moment, precisely how to behave.

So is an effective procedure an algorithm? Before one rushes to answer one should note that although robots and computers might run algorithms, most humans do not do so consciously when they are doing things that are strictly biological or human in nature, like speaking in a human language, yawning, eating dinner, chasing after a missed train

---

[1]Gösta Mittag-Leffler (1846-1927) had an unrelated dispute with Alfred Nobel. Alfred Nobel never included mathematics along with the other categories for Nobel prizes: Physics, Chemistry, Medicine and Economics. Doubtless undergraduates in mathematics have heard of "The Mittag-Leffler theorem," if they have taken a course in complex analysis. The theorem expands certain "meromorphic" functions $f(z)$ on the complex plane $\mathbb{C}$, each as a convergent infinite series of complex rational functions that are analytic on the complex plane except at infinitely many simple "poles."

[2]For more information on this the reader should consult the book *Logic for Mathematicians*, by A. G. Hamilton, and the book *Computation: Finite and Infinite Machines*, by Marvin Minsky.

on a workday, or adding up with paper and pencil, the prices of the grocery items they bought at a supermarket.

Church's lambda calculus made much more rigorous the concept of a recursively enumerable function, a special kind of computable function that inputs a word or string and outputs an integer. If the Turing Machine can use the word input to arrive at a solution for some computable function, then the problem on which it is working by means of some algorithm is *decidable*. Then the Turing Machine can reach a final state, because it solved a problem that was tractable. It will halt in polynomial time, as it turns out. All this means that the Turing Machine can "recognize" the language. Back in the nineteen sixties on TV science fiction shows such as *Star Trek* and *Lost in Space*, both the *The Enterprise* computer and the Robinson family's robot might say, "That does not compute," which for a talking Turing Machine might have been rendered as, "The problem you gave me to solve is undecidable."

A Turing Machine can recognize any recursively enumerable language, which also can be called a *Type 0 language*.

There actually are two types of Turing machines (TM) possible. One is called a *deterministic* TM while the other is a *nondeterministic* TM. The first version becomes important when an algorithm, or set of rules must be implemented in a straightforward fashion, denoted usually as a tape head making various alterations or replacements of words or letters on a tape of infinite length. It must be able to recognize the language represented by the words and symbols on the tape and to do so in what one calls *polynomial* time, where this means in time bounded above by n^k bits, where k and n are positive integers. It is most desirable for the machine to be able to halt, which means simply that things must be done in tractable time so that it does not run forever such as for example, would be the case if it was expected to compute the final digit of a real number like $\pi = 3.14159\cdots$. Srinivasa Ramanujan was able to derive a fascinating way to estimate $\frac{\pi}{4}$. Even so if a high performance computer attempted to compute $4 \times \frac{\pi}{4}$ to the very last digit it would find that the problem is intractable, meaning it cannot be done even with a 'worst case scenario' type of algorithm.

When one considers the various kinds of algorithms that a computer can use to compute the solution to a given problem, Turing also wanted to classify these problems as being either "decidable" or "undecidable." If a problem is undecidable then the machine cannot reach a final or accept state with an algorithm that can run in tractable time. There are several examples of such problems. One was the Halting Problem Turing himself devised. Another undecidable problem was based on a question asked by German mathematician David Hilbert, which we express as follows: *Is there a way to use an algorithm to find a solution to any given polynomial equation in several unknowns and with integer coefficients, that has integer solutions that are relatively prime?*[3] Various mathematicians worked on this problem from the nineteen fifties until the nineteen seventies, including Yuri Matiyasevich, Julia Robinson, Martin Davis and Hilary Putnam. Their results had more to do more with computable and recursive functions rather than working with abstract Turing machines. Yet through their efforts it was shown that Hilbert's Tenth Problem (called *Hilbert's Tenth Problem*, because it was the tenth problem among several other unsolved problems he had posed) is an undecidable problem, which means and relevant to our discussion here, there exists no algorithm that can solve Hilbert's Tenth Problem in a way that will get the TM to reach an ACCEPT state or FINAL state and halt in polynomial time. Martin Davis discusses Hilbert's Tenth Problem and the

---

[3]The type of equation in question is called a "Diophantine equation."

research that burgeoned from it in his book, "Computability and Unsolvability."

Yet Hilbert had posed another problem. Suppose one is given a proposition based on the axioms of first order logic along with some additional axioms. Does there exist some procedure or algorithm that always will prove the given proposition is true or false as the outcome of some sequential process, meaning step by step using the axioms? Alonzo Church and Alan Turing were able to show the answer to this is no.

Still another mathematical problem that was proved to be undecidable where we mean this in the contexts of axiomatic set theory and first order logic instead of in the context of Turing machines, was the continuum hypothesis, which was the first unsolved problem Hilbert presented along with the others in 1900.

In set theory the "continuum" is simply all the real numbers in the open interval

$$(0, 1) \tag{3.26}$$

on the real line, meaning all real numbers x, such that $0 < x < 1$. The "cardinality" (we will denote it here by $|(0, 1)|$) of such an uncountably infinite set of real numbers is some transfinite number, defined by mathematician Georg Cantor. Cantor's system denotes the cardinality of the integers as $\aleph_0$ and the cardinality of the real numbers as $2^{\aleph_0}$ which actually is the cardinality of that set which is the set of all the possible subsets of the set of all integers

$$\{ \cdots, -3, -2, -1, 0, 1, 2, 3, \cdots \}.$$

The continuum hypothesis states that the cardinality of the set $(0, 1)$ cannot lie somewhere between $\aleph_0$ and $2^{\aleph_0}$. In fact its cardinality either must be $\aleph_0$ or $2^{\aleph_0}$, which means

$$|(0, 1)| \in \{ \aleph_0, 2^{\aleph_0} \}.$$

During 1963–1964 mathematician Paul Cohen (1934–2007) at Stanford University was able to prove this problem too is undecidable. One ought to understand however that in this context the word "undecidable" does not mean in the sense of a computer finally reaching an Accept or Reject state after running some algorithm to solve the continuum hypothesis. Rather in this context it has more to do with not being able to prove the continuum hypothesis is true or false by using some axiomatic set theory.

Paul Cohen's result was based in part on a previous result found by the mathematician and logician Kurt Gödel in the nineteen thirties.

Gödel's Incompleteness Theorem makes an important statement that applies to all formal arithmetical systems that are based upon various axioms, mathematical symbols, statements, propositions, etc. Basically it says that one can formulate statements or propositions within such a system that cannot be proved from within the formal system. Actually though we ought to mention that Gödel came up with two important theorems about arithmetic systems, their axioms and the propositions drawn from them. His second result has to do with the fact that a formal arithmetic system can be incapable of proving that it is consistent.

As far as *incompleteness* is concerned however we can give a vivid example. Suppose we use an arithmetic system to create a proposition such as

This Proposition is false.

The proposition asserts that it is false. But if it is false it cannot be a true statement. But then it does state that it is false, which means it is a true statement. So in the first case it is a false statement and in the second case it is a true statement.

Now we all know that Prince Hamlet in Shakespeare's tragedy never proclaimed "To be and not to be," although that might not be true for some quantum theorist or quantum computation expert who also is a dramatist! Likewise "p AND NOT p" cannot be a true statement if either one of p, NOT p is false or if of both of these are false. Let p denote "This proposition is false," and $q = \neg p$ its negation. If one assumes one can assign truth values

$$T, F,$$

for TRUE and FALSE respectively in some truth table for p, q and $p \wedge q$ one arrives at a contradiction, as in the truth table that follows:

p	q	$p \wedge q$
T	F	F
F	T	F

Truth Table for the Liar Paradox.

There is a famous application of the Liar paradox in science fiction that we saw back in Chapter 2 when we considered the android named Norman in an episode from Gene Roddenberry's *Star Trek* television series. Recall that back then Captain Kirk told Norman "Everything I tell you is a lie. I am lying."

When Times get bad for Science

Long past are the days when a rich Rennaissance Florentine like Lorenzo de Medici or a benevolent despot like Frederick the Great would subsidize financially the sculptures of a Michelangelo or the mathematics research of a Leonard Euler. Moreover for the last century funding for research in science, technology, engineering and mathematics frequently is dependent upon government policies and business trends. In Chapter 7 we shall see how this was true even for researchers in computer science and engineering such as Grace Hopper, Claude Shannon and Vannevar Bush back in the nineteen forties. Perhaps it is due to government demands and business trends (that necessarily are profit driven), things that put monetary restrictions on the kind of research that is done, why some software engineers and developers today have organized themselves into independent and small nonprofits to do crowdfunded research in various areas of software engineering and open source development. After all it is very difficult even for gifted research mavericks to do their research together and alone in some small garage while big laboratories run by government, business and academia reap the big millions required to do heavily publicized research.

However to do even this sort of thing, that is, to do research alone and ignored with technical assistants as the Wright brothers or Thomas Edison had done, has not been an option always for other scientists and engineers, as was the case in Europe during World War Two. These were times that were very bad for research in pure science, mathematics and for some forms of engineering like aerospace engineering. Sometimes big government wants fighter bombers and concrete fortresses, not spaceships, lunar colonies and new theorems about prime numbers.

During the war years Alan Turing engaged in important mathematics research on Germany's Enigma code while in 1940 Kurt Gödel, after having been faced with possible induction into the Austrian army two years after the *Anschluss*, left Austria and Germany for Princeton in the United States. Nazism had created a climate of hatred and fear for anyone who simply wished to live in peace, either to do research in mathematics or to sing an amusing song in a cabaret. To the Nazis at the time and to other race prejudiced people in Austria, Poland, Hungary and even in Soviet Russia back in the nineteen twenties and thirties, even brilliant mathematicians and physicists who were Jewish at the time, were considered to be not brilliant at all nor even innovative. For example for the Nazis, Edmund Landau was not an important analytic number theorist but rather a *Jewish* mathematician. Mathematician Issai Schur was not a brilliant group theorist but again, just another "Jewish" mathematician.

Even Einstein in Germany in the thirties was considered by some to be just another "Jewish" physicist, although he is recognized today and rightfully so, for having been one of the most important theorists in physics since Isaac Newton. During the Nazi years in Germany after 1933 and in Austria after 1938 especially after the *Anschluss*, the word "Jewish" in science and in mathematics had become a very demeaning code word or negative qualifier in a society where no Jew whether brilliant, stupid, good or evil evidently, could do anything right for the members of the National Socialist Party and its race mad sympathizers.

Then too good research in both science and in mathematics suffers greatly in a society, whenever blind ignorance and fear, intolerance and ideological extremism replace the reason and moral conscience of an entire nation. In the nineteenth and twentieth centuries before the outbreak of World War Two, Germany and Italy both had given the world much in terms of scientific and mathematical research. Boltzmann, Hertz, Helmholtz, Planck, Pauli, Heisenberg, Riemann, Hilbert, Marconi and Peano are not insignificant names in physics and mathematics. Yet under the Nazis research funding for pure science and pure mathematics went on the wane. Adolph Hitler, Hermann Göring and Admiral Karl Dönitz did not want Einstein's Special Theory of Relativity or Issai Schur's fine results on group representations. No, they preferred faster U-boats, more long range fighter bombers, applied mathematicians who could calculate accurate parabolic trajectories for any artillery shells fired from big guns on the battlefield, and bigger tanks and fortresses. Everyone must conform. Everyone must bend their will to the single cause of the Greater Germany. Many were sucked away down into the maelstrom of this type of "collective groupthink" that was rife within Germany and Austria, including brilliant people like physicist Johannes Stark, the mathematicians Oswald Teichmüller and Ludwig Bieberbach and film director Leni Riefenstahl.

The Nazi effort to exterminate Europe's entire Jewish population is one powerful historical example of how new technologies can be abused in the cause of tremendous evil by a civilization in which a majority of people allow negative and destructive forces like racial prejudice, religious bigotry, nationalism and xenophobia, to get entirely out of control like Pandora opening up her bag of ills. For in January, 1933 after their elections all most Germans could see was a glorious future under the Nazi rule of Adolph Hitler. But in May, 1945, great European cities like Warsaw, Dresden and Berlin lay in ruins and the country of Carl Friedrich Gauss, Beethoven, Schiller, Felix Mendelssohn, Goethe and Richard Wagner found itself under the domination of no less than four foreign nations including a brutally totalitarian communist nation under Joseph Stalin, for almost half a century.

Those who had dominated others suffered the ignominious fate of getting dominated themselves. Luckily for them both Adolph Hitler and Doctor Göbbels escaped all the consequences through suicide. But their people were left with a country in ruins and occupied by foreign armies. After the German Wehrmacht's surrender to the Allied armies under General Eisenhower and the armed forces under Soviet General Georgy Zhukov on 8 May, 1945, it was as if the English poet Christopher Marlowe's Doctor Faustus had made a deal with the devil sealed in blood but again, only to get burned for all his trouble. *Che sera, sera.*

If today after World War Two the human species still cannot evolve to the point at which it knows what it really means for a species to be "intelligent" enough to learn from its tragic mistakes so as to avoid doing them again, then it will be doomed to eventual extinction through self-annihilation. It is very doubtful that any extraterrestrial civilization if they do exist somewhere, will shed alien tears over the irresponsible and senseless behavior that brought the human race along with all its history, to its unnecessary and tragic demise. If one day an alien species elsewhere did happen to learn about the self-destruction of the "intelligent" human race, they might view the distant event as having been Nature's way of preventing an overly aggressive, destructive, irresponsible and far too dangerous species from advancing beyond the confines of their troubled Earth to threaten the stability of other locations in interstellar space.

But in connection with Germany and Nazism, one also must understand the historical context of those times, that led to the irrational, misguided and very costly effort by the Nazis to incinerate all that was among the very best within European civilization from Buckingham Palace in London to Saint Basil's Cathedral in Red Square along with Germany itself even, if they could not have their way with both the world and the Jews.

After 1919 The League of Nations took away Germany's colonies, including in Southwest Africa. Some racially mixed blacks from these African colonies, that is to say, those blacks who had at least one German parent who was either a former settler or plantation owner, lived in Germany after World War I. There were other racially mixed black Germans in the Rhineland[4] whose fathers had been black African soldiers from the French colonies in Africa, mostly Senegalese, and whose mothers were German women. When the Nuremberg racial laws were enforced throughout Germany after the 1936 Berlin Olympics, all German Jewish people, "Gypsies" and German blacks no longer were allowed to exercise any constitutional rights as citizens and they had been reclassified already as "subjects" of the Reich back in 1935. There was no determined effort on the part of Nazi Germany to exterminate the country's entire racially mixed, German black population, although the German blacks in the Rhineland were subjected to physical sterilization so they could not have children. But that was not true for Germany's Jews, Roma and Sinti minority populations. Germany had been a country in which antisemitism had existed for almost five hundred years even before Martin Luther and the Reformation, as far back as the Third Crusade.

Combined with empire and authoritarianism, xenophobia, Prussian militarism and rampant nationalism within Germany's Second Reich from the 1870s until World War I, all these negative human forces back then had become fertile soil for the nourishment of Adolph Hitler's ideas, who found himself in the trenches fighting faithfully for Imperial Germany and for Kaiser Wilhelm II at Ypres. On occasion during the Nazi years some racially mixed German blacks were murdered out of racial hatred or sent to concentration camps. Yet even so there was so much more hatred and paranoia that targeted Jewish people and "Gypsies" at the time in Germany that one could argue convincingly that after 1938 it was safer for one to be a racially mixed German black in Nazi Germany than to be a Jew, a Roma or a Sinti.

---

[4]Adolph Hitler complained bitterly in his book *Mein Kampf*, about the so-called *Rhinelandbastarde.*

One thousand years of antisemitism is a long gestation period of sorts figuratively speaking, for the growth of ideological extremism. So long before the ascension of Adolph Hitler to power in 1933, the country and its people had formed very firmly grounded negative stereotypes about Jews. All a Berliner had to do as far back as 1929, was to read in some morning newspaper about one Jewish bank embezzler or racketeer being arrested, for every other Jew everywhere in the world from Berlin to New York, to become guilty by association[5]. That happened whether the "other" Jewish person in question was a mathematician, a humanitarian, a physicist, racketeer, organic chemist, prostitute, embezzler or medical surgeon.

To many Germans a harmless little girl named Anne Frank was just as evil or as capable of evil as was a Jewish bank embezzler. Absurd, you say? Not to a bigot; for any bigot such insane reasoning makes perfect sense. In many civilizations a prejudiced or bigoted majority of people bases the behavior of an entire minority group population on the immoral or criminal behavior of some sample from that minority population. This is why Eric Hofer warned in his book *The True Believer* that when it comes to the rise of mass movements and hate groups who need to find enemies, the fate of a minority group can depend solely upon the adverse or bad behavior of some of its worst members.

Do not forget that many of the Jewish victims who died in Bergen-Belsen were innocent children, including Anne Frank and her older sister Margot, neither of whom had been a bank embezzler or a racketeer. *Ad calamitatem quilibet rumor valet.* On Sunday, September 15, 1963, when the very idea of black citizens in the democratic United States of America having the right to vote infuriated the Ku Klux Klan, the bombing of the Sixteenth Street Baptist Church in Birmingham, Alabama involved the murder of four innocent children who ironically were too young to vote in the pro-segregationist Southern state. Today we know only too well that many of the murdered victims of the Taliban, Al Qaeda and ISIL also have been innocent children. Doubtless those children were murdered too in places like Afghanistan and Syria, without understanding why ISIS members and the Taliban hated them so much.

Delusion not ignorance, is the nourishment of a bigot. Houston Stewart Chamberlain, Paul Joseph Göbbels, Johannes Stark, Oswald Teichmüller and Leni Riefenstahl? These certainly were not ignorant men and Leni Riefenstahl certainly was not an ignorant woman.

Needless to say, within such a frenzied, violent, highly irrational and intolerant society that was Germany from 1919 until 1938, a society in which Propaganda Minister Doctor Paul Joseph Göbbels could feel free in 1933 to burn up the written works of Albert Einstein, Sigmund Freud and Helen Keller while he stood united among Party members and German university students one night upon a university campus at Opernplatz in Berlin (whether those authors were works by "Jewish mathematicians" or not), a free exchange of mathematical or scientific ideas and free inquiry had become impossible.

At such times in human history whenever things like tolerance, freedom and free inquiry are torn asunder before mass movements, collective group thinking, human brutality and patterns of force, frequently also what men and women mean by the use of the word *truth* is rendered powerless in its impact if not meaningless as well. For example Nazi scientists denounced Einstein as an unoriginal, mediocre Jewish physicist who had borrowed ideas from others with The Special Theory of Relativity. Doctor Göbbels

---

[5]What happened to Germany and its people under Nazism back in the nineteen thirties and forties is it seems, a warning that is ignored today. Nevertheless William Shirer gave a masterful historical description of it all in his massive tome: *The Rise and Fall of the Third Reich.*

attacked the music of Paul Hindemuth and Duke Ellington as being "decadent" and "degenerate." After 1933 the National Socialist version of truth had declared that the novels of H. G. Wells and Ernest Hemingway were fit only to be burned in a bonfire. Even a right wing, conservative German nationalist and Adolph Hitler supporter like Pastor Martin Niemöller became an enemy of the people after he resisted the Nazi Party's efforts to dominate Germany's Protestant churches with ideas on Aryan racial supremacy. When Pope Pius XI, in his encyclical *Mit brennender Sorge* denounced nationalism and racism as evils that were contrary to the laws of nature and to the teachings and spirit of Christ, Doctor Göbbels had the encyclical banned from being read on German radio and within Catholic churches so as not to spread any "lies" about the Reich.

Quid est veritas? That is what Pontius Pilate asked Jesus the Nazarene, evidently not in Latin as in the Vulgate but in the *koine* Greek of his time. Today one popular saying is "Speak truth to power." But in any totalitarian state the actual meanings for words like *truth, good* and *evil* are determined neither by Bibles, Torahs, ecclessiastic encyclicals nor by actual facts and evidence, but by the angry or fearful human mob that wields the biggest axe. It was in fact an angry mob that clamored for the Nazarene to be crucified, not the Ten Commandments or a Roman tribune.

Back in 1940 Nazi Germany, the Soviet Union and even the US Jim Crow state of Mississippi and the Union of South Africa under *apartheid* had their own legal systems and their own definitions for words like "truth," "falsehood," "good" and "bad." One cannot speak truth to power in a society in which power itself determines what is true.

It is no wonder then that Kurt Gödel, whose important work on first order logic, the axioms of arithmetic and set theory required not an axe, but clearly defined mathematical descriptions for truth and falsehood, left for America where he was accepted as a researcher at the Institute for Advanced Study at Princeton and where he became a good friend of Albert Einstein. Some others left Germany and Europe too (especially after 1939) among the more fortunate few within the mathematical, scientific and various artistic circles throughout Germany and Europe, such as the mathematicians Emmy Nöther, Stanislaw Ulam, John von Neumann and Issai Schur, the physicists Leo Szilard, Enrico Fermi, Hans Bethe, Edward Teller and Lise Meitner, the actors and actresses Peter Lorre, Paul Henreid, Conrad Veidt, Hedy Lamarr and Marlene Dietrich, the film directors Billy Wilder and Fritz Lang, composers Paul Hindemuth, Erich Wolfgang Korngold, Bohuslav Martinu, Arnold Schönberg and Alban Berg, pianists Rudolf Serkin and Artur Schnabel and music conductor Bruno Walter, who had been a student of Gustav Mahler. National Socialism might have preached racial purity and Aryan race supremacy, yet the unreasonableness, violence and intolerance of the "racially superior" Nazis forced many of Germany's native born geniuses in mathematics, in science and in art out of the country whether those particular individuals among this diverse group of people were Jewish or not.

On the other hand still another frenzied, violent, highly irrational and intolerant society at the time was Stalinist Russia, from which the theoretical physicist George Gamow and his wife managed to defect in the nineteen thirties. The brilliant Russian pianist Vladimir Horowitz also when faced with a situation of convenient serendipity, that is performing outside his country, managed to defect from the Soviet Union.

In every civilization that ever existed or will exist on this planet, it turns out that human intolerance too, is predictable, just as are taxes, voting patterns (usually), death and robotic kinematics.

But is all this discussion about the Nazi tyranny and about how scientists and mathematicians suffered under it nothing more than mere moralizing about history? No, not when new technologies enter the picture. From 1939 to 1945 Germany did more than to murder Jewish people by the millions across Europe. The Nazis used the latest state of the art military technology at the time, to slaughter millions of others across Europe as well, from blond Anglo-Saxons in London to blue-eyed Poles in Warsaw. Their very irrational and nonsensical hatred of Jews led to their forgetting apparently, that these other non-Jewish victims outside of Nazi Germany were supposed to be "Aryans," according to the racial supremacy theories of "Aryans" such as Houston Stewart Chamberlain, Heinrich Himmler and Alfred Rosenberg. The Nazis accomplished all their unrestrained pillaging, destruction, mass murder and horror across European civilization with the best technology that was available at the time to the totalitarian German state: Enigma coding machines and Panzer tanks for Field Marshal Guderian, Messerschmidt BF 109 planes and Stuka dive bombers for Göring, massive battleships like the *Graf Spee*, the *Bismarck* and U–Boats for Admiral Dönitz, massive, bomb proof fortresses that never were far from England's coast, buzz–bombs, ballistic rockets and also Zyklon B gas in the death camps. Even if one is not a Jew one at least should have the wisdom or just common sense to realize what can happen whenever the citizens in any civilization with access to brand new technologies allow both their bigotry and unrestrained passions to overcome good ethics, decency and sound judgement. *Those who fail to remember the past are condemned to relive it.*

So 'all this talk about the Nazis' is not as irrelevant as it seems to the discussion about robots, certainly not if robots in the far future have at that time the intelligence level of humans, but are contaminated by the flawed behaviors of biased and prejudiced, human men and women. Or perhaps robots one day might become so intelligent they begin to mimic or to emulate humans when humans are behaving at their worst.

It is very possible that one day military robots could be built that can fire automatic weapons and toss hand grenades. Indeed it is possible to build them through mass production, just as Henry Ford mass produced his Model T.

A future world of mechanical humanoid Terminators is not impossible. For passionate and unrestrained human egotism, prejudice and bigotry to become energized into acts of bullying, online "trolling" and violence is one thing. But does one want intelligent machines in the future to learn how to emulate or to mimic this sort of negative and destructive *human* behavior? Even if they are not taught to mimic human emotions, intelligent machines in the future still could learn to do so simply by observing the actions of human beings of the worst possible kinds.

It was the human Doctor Frankenstein who created the destructive Monster, not the Monster who had created the human.

This issue we will address again later, when we consider the possibility that machines can "out think" humans as well as mimic our emotional behaviors whether these are good or evil. Still let us not overreact, because the real danger is *not in the further development of robots, artificial intelligence and intelligent systems.* No, rather the real danger is in creating these things with the intent to get highly intelligent machines to emulate or to imitate human behaviors without providing any effective safeguards or cyber security mechanisms to prevent these new technologies from going horribly wrong or from being abused by evil human beings. That perhaps ought to be avoided.

The mastering of language by machines though, need not be fraught with danger. Even today some machines at least have been developed to deal with the most basic fundamentals of human language. But to do this more effectively requires a better understanding of how to get a machine to deal effectively with conceptualized language, or languages that exists in an abstract or general sense.

A Turing Machine deals with language effectively in an abstract sense and this enables some computers today to emulate human conversation, as we see in the next chapter and later in Chapter 12.

Chapter 4

Turing Machines that Converse

Just how successful have computers and robots been so far, when it comes to holding a conversation with a human in such a way that, if the machine was hidden from view, the human could not tell that it was a machine? Even today some robots like Asimo and virtual humans (chatbots) even, seem to be very adept when it comes to having a brief conversation with humans. But do these machines truly understand the conversation as does a human? Not at present. Even so there are cases when computers have shown themselves to be very capable of uttering verbal responses that seem to be direct and intelligent responses to a human question or statement. In his book, "Broca's Brain," planetary astronomer Carl Sagan gives an account of a hypothetical, human to computer conversation that was devised by Turing, a conversation in which the Interrogator is a human and the Witness is a machine in communication with the human through a teletype system. A second human and computer conversation example he relates in the same book, actually did occur at MIT between artificial intelligence researcher Terry Winograd and a computer with a program dubbed SHRDLU. SHRDLU was a computer, virtualization and computer graphics system Winograd used to allow various virtual, three dimensional objects to be moved around in a virtual scene, so that the experimenter could ask the machine various questions.

One conversational encounter Sagan relates about in his book *Broca's Brain*, and that occurred between Winograd and the computer system is impressive. The human asks the machine when did it perform a certain action and the machine tells the human when it performed the action. Winograd asks then why did it perform this action and the machine explains why it did so[1].

Winograd used Lisp to develop some of the code. This truly was a remarkable and highly successful effort at human-computer interaction and a successful experiment in natural language understanding. The software can be given sufficient modifications to bring it up to date and with some of today's machine learning algorithms included, to give today's generation of machines more adaptive ability for its responses.

4.1 Computer Evolution

Before computers and robots could hold down a conversation with humans with such adroitness that the humans could be fooled into thinking it was talking to another hu-

---

[1]The actual conversation between human and machine is in the Sagan book on page 242, Chapter 20, "In Defense of Robots."

man however, these machines would have to attain much higher levels of technological sophistication than the technology levels of the nineteen forties and fifties when Church and Turing did their research.

Promisingly though, the large computer systems in operation in the nineteen forties and fifties did manage to fascinate one with their computational abilities, the much simpler machines such as the Difference Engine and Analyzer that had been built by Charles Babbage almost one century earlier.

Back in the forties and fifties most of these gargantuan computational, analog machines either were located at sprawling university campuses or in government facilities. One example of such a machine during the years of World War Two was the electromechanical Mark I computer at Harvard University, which also was called the IBM Automatic Sequence Controlled Calculator (ASCC). This was a stupendous powerhouse of a computational analog machine, and the vision of a mathematician named Howard Aiken. Aiken incorporated some of the ideas of Charles Babbage into the working design of the machine. Sprawling more than fifty feet in length and comprised of hundreds of thousands of switches, relays, dials, knobs, gears, shafts and other component parts, this machine could perform arithmetical computations in seconds and minutes, which today one admits seems inordinately long on today's platforms, yet this was not the case for state of the art computer technology more than seventy years ago. IBM's Doctor Thomas J. Watson was instrumental in funding the project.

Between 1942–1944 and with the Mark I at Harvard Aiken had a highly talented group of mathematicians and others working with him. One of these exceptional people was US Naval Reserve Lieutenant Grace Hopper, a mathematics professor from Vassar. She joined the Navy's Bureau of Ships Computation Project at Harvard after graduating in 1944 from the US Naval Reserve's Midshipmen's School in Massachusetts.

In particular the Mark I was assigned the task of solving various numerical physics problems that John von Neumann needed to solve for the Manhattan Project. But it also was used to solve certain ballistics related problems.

After the war Hopper had another daunting task set before her. At the Remington Rand Corp. she had to design a compiler for the UNIVAC[2], so it could accept programs written by the human programmer. Long before Java bytecode and the Java Virtual Machine, many computers needed compilers to resolve programs written by human programmers into a machine language that the computer could recognize[3]. This is true even today for many different, high level programming languages, such as Fortran 90 or C/C++. While a human programmer or software engineer might understand words and terms like **public, private, String, main(), scanf(), read**, etc., most computers today understand only long strings made from digits like 0, 1 and from hexadecimal based digits, stored in memory registers. In a sense even a compiler is something of a robot called a finite state machine. Its task is to recognize certain languages that computer scientists call *regular languages*, a term we shall encounter in a later chapter about Turing machines.

In the two decades that followed the end of World War two, computer scientists entered into a brave new world of computer programming, computer design and data processing. Computer technology and software applications evolved at a very rapid pace.

One cause of this rapid advance in computer science, information theory and informa-

---

[2]Many early computers had names that ended with the letters VAC, like the UNIVAC, or the ENIAC on which John von Neumann worked in the fifties, etc.

[3]Today programmers do this with low level languages, like assembler

tion technology was due to the launching of a small satellite called Sputnik. We discuss these developments in Part II.

Part II

Robots in Matter

Chapter 5

NASA's Space Probes

NASA designed and built literally hundreds of space probes after its formation, space probes to study Earth weather and climate, the solar wind and the properties of the various inner and outer planets in the solar system. At the beginning probes like Explorer 1 and Explorer 2 did yield data of some value yet the technology at the time was very limited. This situation changed however with advancements in computer designs, computer technology, computer software and in solid state physics, in particular with advancements in semiconductor technology such as the transistor and the diode. Then the space probes grew ever more robotic in nature, since they became more and more either like semiautonomous or autonomous, technological remote sensor systems.

5.1 Sputnik 1, Eisenhower, and Explorer 1

The world was stunned then enthralled, on and after October 1957 during the International Geophysical Year, when Soviet scientists and engineers launched the first artificial satellite into low Earth orbit around the planet. This spherical object less than two feet in diameter and not quite two hundred pounds in weight circled the Earth for ninety two days, emitting a radio signal roughly between twenty to forty Megahertz which allowed the signals to be picked up by eager and dedicated ham radio operators. Today in retrospect this might seem to be technologically insignificant. This was not true however in 1957. Back then many people in the United States and Europe could remember the jet powered V1 buzz bombs and V2 rockets that had been designed and developed for Germany at Peenemünde by Doctor Wernher von Braun, during the last four years of World War Two.

For years as a citizen of Germany between the two world wars, Wernher von Braun had nurtured a lifelong vision of space exploration through rocketry. Unfortunately for him from 1939 until 1945 the aims of Adolf Hitler and the National Socialist Reich was winning the war with England, France, Russia and then after December 8, 1941, the United States of America. So instead of designing and building rockets for space travel von Braun had to use his unique talent for rocket engineering to design and to develop an automated ballistic missile weapon against England. However this costly Reich military program, which by the way had many launch failures before any successful launches possibly due either to poor reliability engineering or to ineffective risk management, turned out to be an exercise in futility when it came to helping the Fuehrer win the war. For one reason the V1 buzz bombs and V2 rockets, although quite the achievement for rocket science and automated guided missile flight, were as weapons not manufactured in enough numbers to turn the tides of the war in Germany's favor.

Von Braun surrendered to the US Army in 1945, preferring to surrender his secrets on rocket science to the USA instead of to Joseph Stalin and to Soviet Russia's military engineers, since in many respects the Soviets under Stalin could be just as violent and as militaristic as were Hitler and the other tyrants who helped Hitler to maintain his Nazi regime.

In the early nineteen fifties von Braun was living in America, developing the Redstone rocket for the US Army in Huntsville Alabama. Since he was doing rocket and guided missile research for the military von Braun did not have much time to pursue research in the exploration of space. That changed significantly however after October 1957, that is, after the Soviet Union launched the first artificial satellite.

Sputnik 1 turned out to be a public relations embarrassment for President Dwight David Eisenhower, who as a general had been the Supreme Commander of the Allied Forces at Normandy on June 6, 1944. The industrial nation that had been the biggest winner of World War Two it seemed had become the biggest loser at the dawn of the Space Age. Yet Eisenhower recovered quickly from the abrupt shock of the Soviet Sputnik launch. He compensated by seeing to it that American scientists would make important contributions to much research in meteorology, geophysics, solar science and in other sciences during the International Physical Year. Moreover he put scientists and rocket engineers to work on space satellite technology.

One only can imagine how much this must have gratified Wernher von Braun, who under President Eisenhower became the director of the National Aeronautics and Space Administration in July 1960. Almost three years after the Sputnik launch Von Braun had become the director of NASA's manned and unmanned missions. Yet Eisenhower's response to the two Soviet Sputnik launches was so determined and immediate that the President wisely did not wait for von Braun to start his space missions after 1960.

On January 31, 1958, NASA launched a Jupiter rocket to put the Explorer 1 artificial space satellite (See the Wiki link for Explorer 1 in the Preface) into orbit around Earth with an orbital period slightly less than two hours. Explorer 1 was built at the Jet Propulsion Laboratory. The Explorer 1 payload included mercury batteries and remote sensing equipment which included a Geiger counter, thermal sensors, acoustic sensors to detect micrometeor impacts and an electromechanical sensor also to detect the micrometeor impacts. The remote sensors were designed and developed by Doctor James Van Allen and Doctor George Ludwig at the University of Iowa. The mission was a success not just in terms of its launch but also for the Eisenhower administration's response to Sputnik. Whereas Sputnik did glean some information about the ionosphere due to this atmospheric region's effect on radio signals, Explorer 1 made a truly historical discovery of enormous significance for particle physicists. Explorer 1 sensors had detected radiation belts, caused by high speed atomic particles suspended within Earth's magnetic field. Eventually they came to be called the Van Allen radiation belts.

But were Explorer 1 and its successor Explorer 2 robots? As unmanned probes and artificial satellites they had no ability at decision making. Once Explorer 1 was in orbit it no longer required physical contact from humans to keep it running because it had its own battery supply. It had antennae by which it could transmit its sensory data back to ground stations. It did not run any artificial intelligence or machine learning computer programs for its operations. Yet the use of real time computer programs that run within a space probe to help the machine to automate at least some tasks eventually would become a part of the technology of NASA's future space probes.

5.2 Mariner 9

One of no less than nine probes in the Mariner program, Mariner 9 reached Martian space in November 1971 after a six month trip.

No other robotic space probe bore for NASA the actual distinction of orbiting the planet Mars before Mariner 9 had achieved this marvelous feat and it transmitted more data about Martian topography than did any probe before it. In fact with its sophisticated visual imaging technology it was able to map out most of the planet's physicals terrain for the benefit of planetary astronomers: Mountains, valleys, dry river beds, dead volcanoes. The probe also sent back crucial data about the Martian atmosphere, weather and dust storms on the surface. These dust storms can be very fierce and they pose a serious problem for robot landers on the planet depending upon when and where the dust storms occur and where the rovers land on the surface.

Mariner 9 utilized infrared and ultraviolet spectrometers for atmospheric and temperature studies, solar panels, video cameras, attitude control and a maneuvering engine to orient itself and for navigation in deep space. The digital images were stored in grayscale which can range from eight bits of grayscale to forty-eight bits of grayscale. For bit error correction it used a Hadamard algorithm.

5.3 Pioneer 10 and Pioneer 11

The Pioneer 10 unmanned robotic space probe, launched by NASA and conducted and directed by the Ames Research Center, made its way toward the asteroid belt, the outer planets and beyond the solar system in March 1972. When compared to the more limited technology of the Explorer missions the extensive and elaborate remote sensing equipment Pioneer 10 carried along with the technology it used to manage its telemetry and error correction data transmissions were all far more superior than those in the prior space probe missions. Evidently the reliability engineering and hazard rates for the different technology used in the space probe allowed it to keep transmitting data successfully for at least thirty years, from March 1972 to January 2003, this when it still was sending radio signals from a distance of no less than eighty Earth orbits away from the Sun! This alone was a superb feat of engineering. The probe also was able to determine its orientation to help it to navigate through interplanetary space by using an inertial guidance system that helped to orient itself in reference to three fixed coordinate points, two determined from the fixed sun's position and one from the position of the fixed star Canopus.

Pioneer 10 had a parabolic dish for radio communication and for transmission of various sensor data to and from Earth. The probe had to travel distances that spanned millions of miles away from Earth. This meant that the transmission signals from Pioneer 10 sent back to Earth would be weakened from things like radio wave attenuation or fading. To handle this NASA communication engineers used error correction coding algorithms and "convolutional encoding," so that the robot probe managed successfully any bit error. The probe had decoding technology in two command decoders, memory units to store remote commands but limited data processing and data storage ability (i.e., much less than 10 kilobytes). However still NASA engineers found it necessary to issue commands from a distance.

Pioneer 10 carried along with it very extensive remote sensing technology. It possessed sensors to measure and to analyze the properties of Jupiter's magnetic field. There were probes to analyze the solar wind levels in the outer regions of the solar system. There were more sensors to study Jupiter's radiation belts, sensors to detect meteoroid activity and to measure the density of interplanetary hydrogen and helium. These sensors and the radio transmissions that sent their data back to Earth kept planetary astronomers and solar physicists busy for years.

In many ways the Pioneer 11 mission yielded similar results as did its predecessor. The construction of Pioneer 11, its radioisotope based, thermoelectric energy system, data storage, memory and other technological properties, were similar to those that had been built into Pioneer 10. Yet this space probe did transmit much additional information about Saturn's magnetic field, atmosphere and about Saturn's moons than had any space probe before it.

5.4   Viking 1 and Viking 2

The robot space probe Viking 1 was launched (See the Wiki link for Viking 1 in the Preface) with a Titan rocket on August 20, 1975 and in June 1976 it started to transmit close up images of Mars for the first time. This robot probe truly was remarkable and state of the art back in 1976, for it was equipped not only with sensors to make measurements of weather conditions, temperature, atmospheric pressure, etc. The Viking 1 Lander, which landed in Chryse Planitia, also had the ability to run actual physical and chemical experiments on Martian soil samples to detect any organic compounds present and to test for the presence of microbes or other primitive forms of life.

It would have been groundbreaking for science and an historical milestone indeed for the human species, if the robot Lander had discovered some kind of primitive amino acids! Most of the soil tests yielded negative results, however, except for one particular test, which was positive for the presence of life. Most scientists dismissed this result at the time as a false positive although the discovery of ice on Mars years later has caused some to reevaluate their conclusions.

Viking 2 arrived to enter a Martian orbit in August 1976. The Lander was equipped with the same kind of sensors and soil sample testing equipment that had been fitted into the Viking 1 Lander. Most of the experiments obtained results about the Martian soil similar to those of the previous tests.

Both of the Landers could operate in some ways autonomously and in other ways in a semiautonomous way. By that I mean they could perform sometimes through remote human control. By this the human scientists and engineers back at Earth at least in some way had circumstances almost as good as being physically on the surface of Mars although separated by a vast distance, since the robot probes were in a sense a remote extension of human actions: Press a series of buttons on a control console back at JPL and moments later the Lander digs a trench. Press more buttons and the Lander makes chemical studies.

5.5   Voyager 1 and Voyager 2

As of this year the Voyager probes, both launched in 1977, have gone far beyond all other previous NASA space probes. Soon they will arrive into the Oort cloud and even further onward and away from the solar system, then outward and away into interstellar space. Along with this fact the Voyager 1 and Voyager 2 space probes were extraordinary for other numerous reasons. For one thing both missions sent back digital camera images with the most exceptional resolution of Jupiter, Jupiter's moons, Saturn, Saturn moons, Uranus, the rings of Uranus and Neptune (with help from the probe's Computer Command Subsystem) due to the advanced bit encoding-decoding, error detection and error correction algorithms used (i.e., convolutional coding along with Reed-Solomon codes, Golay codes and Viterbi decoding, as needed) to transmit the image data back to JPL. Various remote sensors enabled them to take close measurements of Jupiter's radiation belts and the radio transmissions of Jupiter and Saturn. They detected a volcano on Io, one of Jupiter's moons, and ice and a possible underground ocean on Ganymede, another one of Jupiter's moons. Voyager visual imaging data identified a bizarre form of hydrocarbon precipitation on Titan, one of Saturn's moons, along with the possible existence there of large bodies of hydrocarbon lakes or oceans. Both probes detected the solar wind in a region far remote from the inner planets and the sun to update what was known about the actual extent of the heliosphere.

Many of these space probes such as Mariner 9 and Mariner 10, Voyager 1 and Voyager 2, in particular those probes that had some sort of on board real time operating system installed or else were teleoperated by some remote ground control, can be called "finite state machines." There were on board feedback and control systems and software that responded to remote input signals from Earth that enforced the software to execute various commands to output changes of state from one state to some other state. We shall see this later when we consider the Mars Pathfinder mission.

Is it possible that either Voyager 1 or Voyager 2 will be discovered still soaring outward somewhere not far from Proxima Centauri, by inhabitants inside some Type I or Type II alien space vessel millions or billions of years from now? If so they will get a very good look at the sort of automated technology that *homo sapiens* had developed before the close of the twentieth century.

5.6 NASA's Mobile Robots

5.6.1 Space Probes

Many of the robot space probes that NASA used for its Mars Science Laboratory Mission were indicative of what it means for a robot to be autonomous. Frequently a robot must perform some tasks autonomously: Navigation, robot motion across an alien terrain, using instruments and sensors to record local mean temperature, climate conditions, etc. Yet on occasion a robot probe such as a probe in low orbit around Earth, might need communication from a source outside its system, such as with a computer or human, in order to resolve some hardware or software emergency or to deal successfully with some sudden or unexpected event or technical malfunction. Technical emergencies also did arise on occasion during some of the Mars space probe missions of the late nineties and in the 2000s, in particular with the Martian Pathfinder and with Mars Spirit.

5.6.2 Mars Pathfinder (With Sojourner Rover, 1996)

The 1996 Mars Pathfinder mission introduced two different types of robot systems, namely the Lander and the Sojourner rover vehicle (See the Wiki link for Pathfinder in the Preface), which for all intents and purposes was a machine in motion on wheels. In a sense this latter machine was a wheeled robot capable of behaving autonomously in a variety of ways, for the most part to enable it to make scientific studies for example of the Martian soil. However the Pathfinder itself developed a serious software glitch that did threaten the success of the mission for at least one day.

The type of real time operating system the Pathfinder computer used had a RAD6000 CPU. Depending upon what tasks or jobs that must be completed by a computer operating system, there can be serious system pathologies that can develop. For example several processes can be running at the same time (a "process is a running" computer program instead of just sitting in some directory or folder as a file) where many of them except for one are using CPU resources. Process 1 might be utilizing Resource 1 and waiting for access to Resource 2 while Process 2 is using Resource 2. But suppose while Process 1 is using Resource 1 for a long time Process 2 is waiting also to get access to Resource 1 and cannot let go of Resource 2 until it has access to Resource 1. This type of pathology in an operating system (See Figure 5.1) whether in a robot, a server or in a laptop, is called "deadlock."

The Mars Pathfinder had a similar issue, only the real time operating system pathology it suffered for at least one day was called *priority inversion*. Most operating systems have processes at different levels of priority, such as high priority, medium priority and low priority. Some CPU scheduling algorithms offer CPU resources first to high priority processes; other algorithms like one algorithm called FIFO ("First In First Out") show favor to whatever process "is first in the queue," so to speak. Pathfinder used an algorithm called "Round Robin." This used an "equal opportunity" type of timing rule for all processes to get their needed resources within the same allotted window of CPU "time slice" or "time quantum," no matter what the process priority. It is a little like ten thirsty people encircling the same water cooler, but each has only a fixed time quantum in which to get one small paper cup of water before he or she has to wait another turn to get another cup for another fixed time quantum. However the Pathfinder real time computer was designed also to allow for interrupts, since it was an event driven operating system. This is required because some unexpected event might require immediate handling or mitigation by the system and CPU at once.

Thus some other preemptive process could get necessary CPU resources when needed to handle an event. In this case back to our water cooler example it would mean someone at the water cooler suddenly is interrupted from getting his or her cup of water while someone else around the cooler suddenly is allowed to interrupt the other person at the cooler, to get his or her own cup. Now imagine there are at least three thirsty people around the cooler.

Thus replacing the three thirsty office workers with three processes on Pathfinder and the water cooler by CPU resource time, a medium priority process suddenly took precedence within a time quantum over a low priority process and the high priority process. But this way the high priority process was kept waiting far too long cumulatively in terms of CPU time/time quanta or time slices. It was as if the high priority process on Pathfinder was thirsting or "starving" for resource time it just was not getting. This is what caused the priority inversion. The Pathfinder's real time computer had to reboot the entire system, in a sense just like an event driven system crash reboot on a home PC, network file server with serious latency issues or laptop. Valuable data was lost with no expectation of disaster recovery.

Fortunately to save the mission JPL was able to provide software patch updates remotely via radio signals, to Pathfinder's real time operating system.

Can this sort of thing happen today to a robot that has a real time operating system running on its embedded computer? Absolutely. All robots in actual practice today

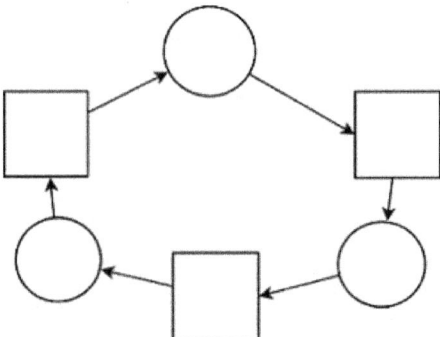

Figure 5.1: Illustration of how computer deadlock can occur. Circles are processes. Squares are CPU resources.

whether as space probes or in industrial manufacturing, just do not look like or act like the more fanciful versions of Tik-Tok, Robby the Robot, R2D2 or Tobor the Great, although Tik-Tok could stop functioning if one did not wind him up each time he needed to function. Today many robots, including semiautonomous and autonomous machines, have running operating systems installed on their computers. So if a robot has a computer operating system, chances are it can experience things like deadlock or priority inversion.

5.6.3 The Mars Spirit and Opportunity Rovers (Launched in June–July, 2003)

These two robot probes, in many ways identical and designed and built to examine Martian rocks and minerals in the soil, test for signs of present or past water activity, take spectrometer readings, photograph and perform other scientific tasks, were launched within a few months of each other and landed successfully on Mars at two different locations. Mars Spirit was a mobile (e.g., wheeled) robot was able to traverse and to navigate successfully almost five miles of Martian terrain. Nevertheless in May, 2009, one of its wheels got lodged in the Martian soil. This is just the type of event in a mobile robot which has motivated and continues to motivate artificial intelligence and machine learning researchers to derive better algorithms and software, to enable a mobile robot being confronted suddenly with a crisis such as a wheel getting struck in soft sand or in mud, to adapt by learning quickly what to do in the situation, to resolve the issue with a quick resolution and one would hope in minimal time.

After months of running simulations, NASA engineers updated the software remotely, but in unsuccessful efforts to dislodge the wheel. Eventually the actual mission ended in early 2010.

Another crisis developed for Spirit from a breakdown in operations when the machine's sleep mode state had a fault. Remote commands were sent to the robot probe by the engineers back home to try to shutdown the rover to get it to reboot to repair the software if that was the issue. What was expected to happen was for the robot probe to repair any software bug by rebooting to resolve the bug, as if it was obeying some code that was similar at least in goal although not in actual fact, to the following illustrative pseudocode:

```
!Boolean fault condition
FAULTCONDITION = 0
!Some more coded lines in here...
WHILE FAULTCONDITION = 0
   REBOOT()
UNTIL
   FAULTCONDITION = 1
End
```

where false and true are represented as 0, 1, respectively. The rover's computer system was expected to reboot to resolve and to correct a software bug. Unfortunately it was looping indefinitely and it was not correcting the bug that way. The robot probe was stuck engaging in an endless cycle of reboots.

JPL engineers deduced eventually the problem was not hardware failure but some software failure in the robot probe's flash memory. More specifically the file management on the flash memory had on it too many unnecessary files. This could pose a problem under conditions of high latency, such as when the probe might be running many processes to help it to conduct several different scientific tests at the same time. So NASA mitigated the problem through bypassing any reboots from the software on the flash memory to reboot the operating system through the system processing of remote radio signals. Indeed something similar can happen even with operating systems that are not real time operating systems. For instance a computer expert at times can reboot a malfunctioning operating system on some home or business office PC with a CD that contains the proper system rebooting files for the system, to avoid reboots from any faulty memory stick or from any hard disk operating system in use. Both Spirit and Opportunity got some remote system software updates in 2007[1].

5.6.4   Mars Curiosity (Launched in November, 2011)

Gradually as time progressed from 1996 to 2011 the real time computer systems that were used on the various orbiters, landers and rovers for the Mars Science Laboratory Mission had upgrades and improvements in computer memory, memory storage capacity and processing power. The later missions included versions of the Electrically Erasable Programmable Read-Only Memory system (EEPROM), which had both a serial bus and a parallel bus and which was able to operate independently from an external power source. These machines also had flash memory and volatile memory (RAM).

In addition to having features such as these, the Curiosity rover had actuated wheels capable of movements independent from each other, as well as a robot arm about seven feet in length. It also had the ability to drill to extract various physical samples from the Martian surface and a soil scooper. The probe examined weather patterns and rock and soil composition. In effect Curiosity was able to do many things a human geologist or meteorologist could have done on the Martian surface, but in a far more economical way than would have been a manned mission to Mars to study the surface, not to mention less arduous and dangerous if dangerous conditions had arisen during a mission such as a serious Martian dust storm.

---

[1]One can find a more detailed description of the Mars Pathfinder computer deadlock (See Figure) matter in the book by Kay A. Robbins and Steven Robbins, *Unix Systems Programming: Communication, Concurrency, and Threads*, Chapter 13 (See Bibliography).

The real time operating system for the rover's computer (called Rover Computer Element, or RCE) had a Rad750 CPU upgrade to the previous RAD6000 CPU that NASA used in the previous Mars Exploratory Rover missions. The RAD750 could handle four hundred million instructions per second in contrast to the previous RAD6000 which handled about thirty-five MIPS. In comparison the Intel Pentium processor in 1994 could handle 188 MIPS. According to roboticist Hans Moravec at Carnegie Mellon a home PC has the intelligence level of around one thousand MIPS, that of a cockroach. Hence it can be said the processors aboard the rovers did not give their computers the intelligence level of an insect!

At the time of this writing one of the most recent autonomous space probes to explore deep space by NASA has been New Horizons, something which has enabled astronomers to get a much better understanding of not only Pluto, but also about several other planetary bodies that lie just beyond the outer part of the solar system.

Later we will discuss further about the future possible MIPS based intelligence levels to measure machine intelligence in a future chapter, where we will discuss the possible types of machines that will exist when their intelligence levels far surpass that of humans.

5.7 NASA's Robot Space Probes and Intelligence: Are Space Probes Safe?

Perhaps first here we should clarify that there can be an enormous difference between an autonomous robotic system and an intelligent system. With most of the NASA robot space probe technology, Mariner 9, Pioneer 10, Voyager 2, New Horizons, etc., these more or less functioned either as semiautonomous or autonomous robotic systems. In fact NASA's space probes were and are still magnificent examples of machine automation. However one should classify these space probes as being more like finite state machines that used logical rule making (YES/NO) to execute automated tasks that were based on such logical decisions. In contrast to that the movie *2001: A Space Odyssey*, gave us the Hal 9000 computer system which was truly an intelligent system. This does not mean Hal 9000 could not act autonomously. It means through comparison that although many of the NASA robot space probes might have behaved autonomously depending on what they were doing, they did not make decisions in ways that simulate or emulate the cognitive human decision making process. So in comparison to the Hal 9000 we can argue quite convincingly that the NASA space probes were safe while Hal 9000 was quite capable of committing homicide and practicing deception and evasion.

But even an autonomous robotic system can be dangerous under certain conditions. Imagine a future major city in which the local government allows commercial businesses to operate autonomous cars, vans, trucks, etc., on the roads at all hours. Some people suggest these autonomous vehicles should connect through the Internet to share data when necessary, such as for the real time updating of traffic routes or destinations. Now add just one very determined and malicious or homicidal human hacker, perhaps working for a foreign power, into the autonomous vehicle network after fingerprinting and mapping it. Authentication protocols are bypassed and confidentiality through encryption is violated. In such a scenario things could get very hazardous indeed, on city roads, streets and even interstate highways. Perhaps there could be even legal repercussions for companies that operate such Internet connected, autonomous vehicles. We discuss matters like these in Chapter 13.

Chapter 6

Cybernetics, Cyborgs and Information

6.1 What is Cybernetics?

Within the first two decades after the close of World War Two there was a sudden rapid growth in five other areas of science, technology and engineering research in addition to space science and technology as we have seen in the previous chapter: Computer science, computer engineering with its rapid increase in data processing power due to the invention of semiconductors (in particular the transistor), robotics, information theory and the new science of cybernetics.

Today many people have the notion that "cybernetics" is some kind of science that deals with combinations of humans and machines. Yet this at best only is a very reductionist description. It is more comprehensive to say that, as a science, cybernetics is the study of any complex system which reacts to input signals from an environment with specific output responses, engages in certain regulatory behaviors that depend upon the kinds of inputs it receives and is capable of both learning and adaptation. More than likely cyberneticists would give even a more comprehensive or detailed definition. Yet the basic idea is a cybernetic system can be any system that reacts to inputs from its environment and then regulates its responses to those inputs. Some examples could include human and computer interaction, such as an employee working at a telephone call center by first interacting with a customer on the telephone (input) while after that he or she immediately updates the customer records on a PC.

Still another example of a cybernetic system would be an inventory and shipment delivery system, where customer names and addresses are updated in a database, where shipments are scheduled through various scheduling algorithms, and where shipment route times are regulated and optimized. There even are cybernetic systems for those systems that are described by and studied within the social sciences and economics, and also in biology, cognitive neuroscience and medicine. The idea is that cybernetics need not involve only technological systems that interact with humans; they can be organizations or large companies that conduct daily business operations in various branches and departments, university systems in which the Registrar's Office staff makes decisions about class schedules then alerts students and faculty, where the professors in various departments order books for an approaching semester and students make dormitory arrangements. But then a cybernetic system also could be a chimpanzee interacting with a termite mound by piercing a hole in the mound with a twig to fetch a six-legged protein snack, or ants in an ant colony scurrying about to make repairs after someone inadvertently steps on the ant mound in some suburban backyard.

6.2   Norbert Wiener (1894–1964)

The year 1948 was significant in politics, in world events, in sports, in music, in literature and in science. That year Mohandas Gandhi, who had been attempting to build bridges of peace between Moslems and Hindus, was assassinated by a Hindu extremist. Chiang-Kai-Shek became President of Nationalist China. The Airlift was America's response to Stalin's blockade of Berlin and around the same time a rift formed between Stalin and Yugoslavia's Tito. After innings the Cleveland Indians won the World Series over the Boston Braves. *The Treasure of the Sierra Madre* took its rightful place as a masterpiece in cinema and in music composer Walter Piston saw his *Symphony Number Three in E* premiered by the Boston Symphony. Literature witnessed the triumph of two literary milestones: *The Naked and the Dead* by the novelist Norman Mailer and *The Pisan Cantos* by the controversial modernist poet Ezra Pound.

The worlds of science, applied mathematics and telecommunications engineering also witnessed two extremely eventful milestones that year: The publication of the book *Cybernetics: Control and Communication in the Animal and the Machine*, by Norbert Wiener and the publication of the paper *Mathematical Theory of Communication*, by Claude E. Shannon.

Some might not see that cybernetics has had some important impact in today's world, except perhaps by definition, meaning one might look at an entire sales company's network of delivery trucks, PCs, inventory software and shipping schedules, while watching the human employees interact with all these things, then say "Okay, so I guess all this is supposed to be a cybernetics system. So what?" So why do I claim these two works by Wiener and Shannon were milestones in science, applied mathematics and telecommunications engineering? Because much that we take for granted today, the technology and various systems in society that we use or interact with on a daily basis, are based upon ideas that were developed by these two researchers, as soon we shall see.

Norbert Wiener, a child whose earliest and original ideas germinated perhaps when he was subjected as a child to home schooling provided by his father, grew up to become a Renaissance man, of sorts. He received his Bachelor of Arts in mathematics at the age of fourteen. One can compare this intellectual accomplishment to that of the English mathematician and physicist William Rowan Hamilton, who could read no less than ten dead and living languages by the age of twelve. Also in comparison Enrico Fermi had solved a boundary value problem, along with its partial differential equation for the vibrating rod, at seventeen. Across the whole span of his active life Wiener had produced a prodigious collection of refereed papers, articles, books and other writings on first order logic, stochastic processes and time series, Brownian motion (a new stochastic process in fact was named after him, called a *Wiener process*), harmonic analysis, cybernetics and even in zoology. He also was very interested as to how cybernetics could be used to solve some of the world's most perplexing and disturbing social and international problems.

Much has been written already by other authors about the absentmindedness of Norbert Wiener. I do recall an anecdote related about Wiener by a physics professor whose class in modern physics (i.e., special relativity and quantum mechanics) I once attended back in 2001 when I was an undergraduate major in mathematics (I also should add here that computer science was my minor) at the University of Massachusetts Boston. Our professor had related that Wiener preferred studies in mathematics to studies in physics and how someone claimed that on his street in Massachusetts near his home residence in Massachusetts (when he was a professor at MIT) one of his children had to stand near the family home to remind him where it was located.

During World War Two Wiener served the US Army by working on some challenging problems in antiaircraft gunnery and ballistics. The main problem was how to automate an antiaircraft gun along with radar, so that an enemy plane could be tracked accurately for it to be targeted successfully by gunfire.

From the radar we can determine the sequential positions of the plane in seconds or in minutes. Since this statistical data is taken at times that are both sequential and successive, the data taken this way usually is called a "time series." One happens to know in advance the actual stochastic process in question. The idea is to use some method to estimate this stochastic process with minimum mean square error. Assume the plane is not banking so that it changes either its altitude or velocity due to acceleration. The radar sightings allow for a sequence of velocities or positions with errors to be plotted where each of these is at a different time interval each time, one second, two seconds, or one minute, two minutes and so forth. One gets the best computational results with the most recent data, since there is the possibility the plane might change its course. Wiener developed a special applied mathematical tool called a *filter*, by which the possible velocity or position estimates for the plane could be predicted with only a minimum of mean square error. For those readers with some familiarity either with linear algebra or with the kinds of vectors that are discussed in classical physics courses that cover statics and mechanics, the data set found by the radar had weighted values added to it, in a way similar to what one does when one expresses a linearly independent vector as a linear combination of much smaller, linearly independent basis vectors. For this reason the filter Wiener devised is called a linear filter.

The Wiener filter, which is meant to be used for systems called "linear systems" (control systems engineers in the fields of electrical engineering and mechanical engineering know only too well what this term means) proved to be highly useful for many researchers across several disciplines in science and engineering including in electrical engineering and mechanical engineering, not only in ballistics. Shortly after Wiener had derived this result the Russian mathematician A. N. Kolmogorov came up with a linear filter for data taken from discrete sets of numbers, since Wiener's result applied in the continuous case, that is for example, where the mathematics and data involved required the use of analog machines. Kolmogorov's version of the filter is very amenable to computation by digital computers. However there also are other kinds of nonlinear filters such as the Kalman filter, to deal with time series data from nonlinear systems (e. g., an inertial navigation system) that cannot be treated very well with Wiener's linear filter.

The Wiener filter is useful for stochastic processes that are "stationary." A stochastic process, sometimes denoted like $X(t)$, is a random variable for each time $t \geq 0$. When the times are discrete integer values $n = 1, 2, 3 \cdots$ then $X[n]$ is called a "stochastic sequence." A stochastic process and a stochastic sequence have some kind of probability distribution. So if we suppose $x[n]$ represents the actual signal (as a real number) one expects to see while one also finds an estimate $y[n]$ derived from the time series in the data set and where $y[n]$ is the input into the linear filter, one needs to minimize the mean square error that one gets from using $(y[n] - x[n])^2$.

Wiener's result enabled engineers to use the linear filter on "stationary" stochastic processes. Since the definition for this is precise, accessible and understandable usually by those familiar with mathematical topics like stochastic signal processing and control theory, I will give an example.

Suppose $X[n]$ is a stationary random sequence. Then what we mean by this simply is we do not expect the signals to vary over time so that the probability distribution changes. For instance if

$$X[1], X[2], X[3], \cdots \tag{6.1}$$

is to have a specific probability distribution, then we should expect that

$$X[1+5], X[2+5], X[3+5], \cdots \tag{6.2}$$

to have the same probability distribution, and that in general,

$$X[1+k], X[2+k], X[3+k], \cdots \tag{6.3}$$

also should have the same probability distribution, for any fixed integer

$$k \in \{1, 2, 3, \cdots\}. \tag{6.4}$$

In fact if n, k, are two natural numbers, then if the two discrete random sequences $X[n]$ and $X[n+k]$ both have the same kind of probability distribution, then $x[n]$ is called a *stationary* stochastic or random sequence. In the continuous case one calls $X(t)$ a stationary random process if both $X(t)$ and $X(t+\tau)$, where $0 \le t, \tau \le \infty$, have the same probability distributions.

So what do all these linear and nonlinear filters for stationary, stochastic or random processes or sequences have to do with cybernetics, one might ask? *Plenty*. These applied mathematical tools were developed precisely to help to implement feedback and control into some of the very kinds of systems that Wiener had characterized as being cybernetic systems, such as various physical systems that use technology and the laws of classical mechanics. The various robot space probes and Martian rovers we learned about in the previous chapter serve as examples of physical systems that utilize error correction for feedback and control. So they do comprise in a sense cybernetic systems. Curiously one might ask why Wiener's work on cybernetics, as well as the Kalman filter, are of any importance when the importance is all around us, whether the system in question is a mobile robot navigating its way across the alien Martian landscape, an antiaircraft gun that shoots down an enemy aircraft, a sprinkler system activated in a building when the internal environment gets too hot, or a robot space probe that uses an inertial navigation system to determine its current position and orientation along three coordinate axes in space.

Are there ways to utilize principles and methods from cybernetics that might accomplish more than the development of inertial guidance systems and the minimization of mean square error in ballistics problems? There are some cyberneticists who believe it can be used to solve some of our world's most vexing and serious domestic and international problems, including overpopulation, disease epidemics like the Ebola outbreak in 2014–2015, terrorism, violent crime, poverty, food shortages, unemployment, war, race prejudice, xenophobia and ethnic unrest.

It is conceivable some feedback and control mechanism could be designed to regulate financial trading automatically in a way that minimizes any potential long term damage in the overall economy from sudden turbulence in the stock market that leads to for example, an economic recession, depression or hyperinflation, perhaps by enforcing some kind of automatic financial governance or protocol once things go wrong. There might be ways to implement feedback and control mechanisms in urban planning that will minimize aberrant human social behaviors, such as violent crime, racial tension and rioting, *while at the same time* maximizing citizen freedom, liberty and security for all without human biases to encroach as destructive factors. If one day someone can derive from the science of cybernetics, some useful set of algorithms to enforce various feedback and control mechanisms within human societies and that is capable of utilizing principles and concepts from social psychology, sociobiology, dynamical systems theory, information theory, linear programming, stochastic processes, decision theory and communications theory, to transform our American society in particular and our present world civilization in general into fully stable societal systems with minimal chaos and antisocial behavior, then cybernetics would be able to accomplish something that no politician on the political left or right and that no religious dogmatist anywhere as well, has been able to do within the last several millennia.

We mentioned there were two important contributions in science and technology in 1948 of overwhelming significance. We already have seen the development of cybernetics and Norbert Wiener's mathematical methods as well as those by A. Kolmogorov and Kalman that helped to create today's linear and nonlinear feedback and control systems. Yet another equally important milestone were papers by Claude E. Shannon, whose research spurred a veritable revolution in electrical engineering and electronic communications.

Most graduate students regardless of their chosen disciplines but particularly in STEM programs (Science/Mathematics/Engineering/Technology), are required to do extensive, independent research, not only to find answers to their own questions, but also to complete research papers assigned by professors (e. g., such as in neuroscience, physics, computer science, information technology or in one of the engineering disciplines) and in preparation for future research endeavors. Frequently as well such an approach to one's graduate education and training is crucial in preparation for a Masters dissertation or capstone project (with this latter option frequently being the case in graduate information technology programs) before the awarding of the Master of Science degree. Seldom however does a Masters dissertation or capstone project rise to the level of extremely high significance in the area of academic research.

This has not been true of Claude Shannon.

Actually Claude E. Shannon (1916–2001) wrote not one but three highly significant papers that have relevance not only in the fields of cybernetics and robotics, but also for telecommunications, digital computers, satellite signal transmissions, machine learning, artificial intelligence and even encryption systems.

6.3 Information Theory, Cybernetics and Robots

Claude E. Shannon had the distinction of being acknowledged for having written one of the most important Masters dissertations from the beginning of the twentieth century until the present generation. In 1948 he also wrote a paper that had an enormous impact upon electronic communication systems, whether these are telegraph or telephone systems, radio signals, fax or even the digital based communication systems which we use today, including modern communication by smart phone, instant messaging and email.

In fact no less than two of Shannon's papers on noiseless and noisy communication channels have become seminal papers, responsible for the astounding growth of applications in digital information technology from them in data communications, data storage and data compression.

One can say that Shannon enjoyed a full and rich life filled with the intellectual gems of technical ingenuity, invention and discovery as well as research in applied mathematics. As a boy growing up in Gaylord Michigan he put together various mechanical devices with ease, built model airplanes and experimented with wireless telegraphy. Doubtless this latter skill enabled him years later to formulate or to design within his mind technical conceptions of his mathematical results. Shannon pursued a sort of dual academic track, so that he obtained undergraduate degrees both in electrical engineering and mathematics, at the University of Michigan.

In 1936 Shannon was a graduate student in electrical engineering at MIT. His mentor at the time was Vannevar Bush (1890–1974). Bush from the Roosevelt years of World War Two and even until today has had an impressive reputation for having spurred new research and development on some of the most important and monumental scientific and technological innovations in the history of the human species. Under FDR Bush had been Director of the Pentagon's Office of Scientific Research and Development. Also in the forties Bush was the chief scientific supervisor and advisor for the Manhattan Project. In this role he served as go-between for J. Robert Oppenheimer and other nuclear scientists whose task it was to confirm "Oppie's" technical models and calculations for the atomic fission bomb. Bush also shared information and news on the project with General Lucius Clay and various politicians.

However while he was at MIT in the thirties Bush had been working on his "differential analyzer." This analog computer was designed and built to a system of electromechanical relays and switching circuits to solve linear differential equations with anywhere from a small number to a large number of independent variables.

6.3.1   The Differential Analyzer and Shannon's Masters Dissertation

Many physicists, applied mathematicians, astronomers, electrical engineers, mechanical engineers, even cardiologists interested in blood circulation and flow and epidemiologists who study the epidemic spread of deadly infectious diseases such as Ebola, have interest in finding solutions or approximate solutions if need be, to various differential equations or systems of differential equations related to their fields. These equations chiefly are used to model mathematically how some physical phenomenon varies over time, space or both. Their solutions yield valuable insights into the nature of various physical systems. For instance when one treats the planets of our solar system as particles in a plane and attracted to the sun at the center, where gravity is treated as a central force $f(r)$ (this just means here a gravitational force in the plane directed radially from the Sun to each planet), one has a certain nonlinear differential equation

$$f(r) = \frac{mh^2}{r^4}\left(\frac{d^2r}{d\theta^2} - \frac{2}{r}\left(\frac{dr}{d\theta}\right)^2 - r\right), \tag{6.5}$$

(where m is the point particle's mass, r, θ are polar coordinates instead of Cartesian coordinates on the plane, and h is an important special constant) by which one can derive planetary motions in the plane as being ellipses or circles, as conic sections around the sun depending on the eccentricity ϵ, described in polar coordinates r, θ. Most electrical engineering students in the earliest years of their studies are exposed to a second order, linear differential equation

$$L\frac{d^2q}{dt^2} + R\frac{dq}{dt} + \frac{1}{C}q = V(t), \qquad (6.6)$$

one uses to model the physical changes in quantities such as voltage (V), resistance (R) and inductance (L) over time. The solution $q(t)$ (which usually is a sum of sine and cosine functions) describes the flow of electric current $q(t)$ in an electrical circuit. Sometimes though the differential equation is not "linear," where this means the changes $\frac{d^2q}{dt^2}$ (second order) or $\frac{dq}{dt}$ (first order) do not appear as values that are squared, cubed or raised to some other positive integer power. On the other hand there are differential equations that model aspects of physical nature but that are not linear in their form, such as the nonlinear differential equation above for the central force $f(r)$. To illustrate graduate students in general relativity might be familiar with Einstein's equation. By the term "Einstein equation" one does not mean the famous well known equation from Special Relativity, namely

$$E = m_0c^2, \qquad (6.7)$$

which (although not a differential equation) relates energy to rest mass and the square of the speed of light and which some in special relativity physics might argue is written more correctly as

$$E = m_0c^2 + O\left(\frac{1}{2}mv^2\right), \qquad (6.8)$$

but rather Einstein's gravitational field equations (we use plural for "equation" since it has multiple solutions)

$$R_{ij} - \frac{1}{2}g_{ij}R = 8\pi G T_{ij} \qquad (6.9)$$

(the quantity R_{ij} with the two indices i, j, is called a covariant, rank two tensor by some physicists) a fascinating nonlinear differential equation from the General Theory of Relativity from which one obtains Einstein's field equations, where g_{ij} is called a metric tensor, G Newton's gravitational constant, where T_{ij} is the covariant stress energy-momentum tensor and R_{ij} is the covariant Ricci tensor (rank two) one derives from a much more elaborate mathematical construct called the rank four Riemann-Christoffel covariant curvature tensor[1], derived by Einstein and from which physicists have derived solutions for gravity waves.

More specifically the Bush Differential Analyzer at MIT in the thirties (which we add here was not the only such analog computer in use) was built to solve certain kinds of linear differential equations. The easiest were linear differential equations of a kind like

$$c_1\frac{d^2y}{dt^2} + c_2\frac{dy}{dt} + c_3y(t) = 0, \qquad (6.10)$$

---

[1] Readers with a good calculus background and with some background in partial derivatives and differential equations too can find sources with more detailed descriptions of these tensors that are used in general relativity and on Einstein's field equations, in the books *Principles of Physical Cosmology* by P. J. E. Peebles, and *General Relativity and Cosmology*, by astronomer G. C. McVittie, although some of the material in this latter book is rather outdated. In the Peebles book one will find an introduction to (nonCartesian)tensors in Chapter 8 and material on the Einstein field equations in Chapter 10 (See Bibliography).

C_1, c_2, c_3 three real constants, with solution

$$y(t) = A\cos t + B\sin t, \tag{6.11}$$

where initial conditions[2] given for the equation help to determine the values for the constants A and B.

The Differential Analyzer was a large (i.e., large enough to fill a large room) electromechanical analog computer that included physical parts to handle numerical tasks like addition and multiplication. One component of the device that truly was ingenious however, was an arrangement of horizontal glass disks and smaller vertical steel wheels with sharp edges and perpendicular to the disks. The disk was mounted on an axis that enabled it to rotate while the perpendicular steel wheel rotated above it, both with angular velocities that had specific values that were crucial to the solution of each particular differential equation at hand, where the angular velocities changed along with the equation to be solved.

The vertical shaft on which the horizontal glass disk was mounted was free not only to rotate the disk but also to move the glass disk horizontally, so as to subject it to both rotation and translation, while the small steel wheel above it was allowed always to move upon the top of it. This movement was similar to the behavior of a phonograph that played a 33 RPM vinyl record, only that in this case the vinyl record and turntable could move horizontally as well as rotate, while the needle above them was replaced with a rotating steel wheel.

One would input values for the differential equation, for example for $\frac{dy}{dt}$, $\frac{d^2y}{dt^2}$, etc. The computer was designed to compute integrals through an incremental and iterative process. This was done by computing the values for both the disk and wheel angular velocities, adding these different values as a sum incremented along with differentials, dt, say, to construct eventually an arc. The end result was an output device that drew a curve that was in effect the graph of the solution. This was done for linear differential equations. If the linear differential equation was second order one added a system of shafts and wheels that were allowed to twist in different directions to represent different values of torque for numerical iteration.

It took incredible skill in the knowledge and use of electromechanical devices and a flair for the inventive, not to mention a thorough understanding on various kinds of differential equations, how they are solved and about the families of parametric curves or surfaces for the solutions (something most undergraduates who majored in mathematics have covered, as in fact I did) for one to build the Differential Analyzer which Shannon used as a graduate student at MIT. The analog computer applied an algebraic system devised by the English mathematician George Boole, in that it used various relays and switching circuits. In his Masters thesis and then later in a refereed paper w3ith the title "A Symbolic Analysis of Relay and Switching Circuits," Shannon was able to show how various relay and switching systems, such as analog computers, telephone networks and even computers, used physical operations that have mathematical models given from Boolean algebra. Conversely he showed also that Boolean algebra can be used to design various physical switching circuits for telephone systems and analog computers.

Yet in 1948 after he had obtained his Masters degree and PhD at MIT, it was his seminal paper "A Mathematical Theory of Communication" that appeared in the *Bell*

---

[2]The more adventurous reader with a solid background in calculus can find definitions for some of the terms used in this Section in textbooks provided in the Bibliography.

System Technical Journal that helped to initiate an explosion of applications in telephone technology, radio, television, satellite communication, wireless communication and the Internet throughout all the subsequent decades since 1948. It just so happened that, more than forty years previously from 1901 to 1903, Guglielmo Marconi had established wireless telegraph communication across the Atlantic.

Wired telegraph communication by Morse code via electrical signals through copper cable, had been fraught with difficulties even immediately after its burgeoning years. For one thing such Morse code communication through underwater cable from country to country had serious problems with electric signal degradation. For instance a dash might be sent for too short or for too long a time. Suppose one wished to wire WAR BETWEEN NORTH AND SOUTH across the Atlantic about the Civil War between the American States. Depending upon how the telegraph operator tapped the keys for the various Morse symbols, the signal for "W" might interfere with the signal for "A," which would create confusion for the overseas recipient as to what symbol was sent. On the other hand a signal for "W" or "A" might be tapped too slowly by the telegraph operator, so that the received signal was so corrupted with noise so as to be unreadable by the recipient.

Years later there were new solutions but also new problems. As John R. Pierce elaborates upon in his excellent, informative and very illuminative book, *An Introduction to Information Theory: Symbols, Signals and Noise*, Thomas Edison devised a way to prevent such signal corruption through "intersymbol interference," by using a "quadruplexing" approach in which he varied current intensity $(1, 3)$ and current direction $(+/-$ for forward/back, meaning dot/dash) to send wired telegraphic signals, so that there were four states possible, that is,

$$-1, +1, -3, +3. \tag{6.12}$$

However even this modification did not eliminate completely the problem of intersymbol interference, depending upon what sequence of symbols was sent.

Others used mathematical models to investigate the problem. From Fourier series and analysis methods mathematicians were able to show how wave attenuation and time delays due to the presence of phase shifts in electric signals transmitted across telegraph wires, can happen when one sees how the signal is comprised of individual, linearly independent wave components. Some of these wave components have frequencies such that the signals suffer wave attenuation, phase shifts or both. When this happens the overall combined signal can suffer distortion. With the use of logarithms to various bases and Fourier analysis, Harry Nyquist studied what things influence the quality of information signal transmission by telegraph. Then in 1928 in a paper he showed that for transmissions restricted to $2N$ electric current signal values per second where N is a large integer, one can receive fairly distortionless signals provided all "redundant" signal frequencies or wave harmonics at N Hz or above (N Hz sometimes is called the "Nyquist frequency") are filtered out.

indent But Nyquist's approach eliminates distortion not only from telegraph signals. It also helps to remove distortion or spatial and temporal "aliasing" from audio, video (television) and digital signals. This also explains why video signals and other data from the NASA robot space probes that transmitted data that had been subjected to various algebraic error correction coding algorithms, could be received back at JPL with little or no signal distortion or aliasing. Other researchers such as R. Hartley, devised useful mathematical results for extending Nyquist's approach to telephone communication systems.

Then in 1948 Claude Shannon published his monumental and historic paper on information theory. In his paper "A Mathematical Theory of Communication," Shannon answered the general question: What is the best way to transmit an information signal across a noisy communication channel, in such a way so as to minimize error?

In order to appreciate the beauty of Shannon's ideas one needs to get a better intellectual grasp on certain mathematical ideas from probability theory, ideas such as random variables, entropy or uncertainty, and noise.

In today's hectic, even chaotic international civilization, we are exposed daily to all sorts of probable events. Our daily weather reports that we glean either from television or from the Internet inform us there is a "chance of heavy rain after 1 PM," "There is a likelihood of a tornado in this county sometime after 8PM," or "Eighty-six percent of those voting respondents taken from a random sample of the population have an unfavorable opinion of the candidate," etc.

Hedge fund investors or other investors might consult financial engineers or Wall Street "quants," to determine how to invest or how to diversify a portfolio. Statisticians in the health or pharmaceutical industry use clinical trials and blind or double blind studies to determine if some specific treatment with a new drug has high significance for some or other numerical value within three standard deviations. They also use qualifiers like mean square error meant to be minimized, as N. Wiener had done in his work. All these mathematical and computational approaches indicate all events do not occur from physical laws of absolute determinism, or with probable outcomes that either are zero or one in value. If someone drops a sack of flour five pounds in weight down to the ground from the roof of a building one hundred feet in height, one can predict when it will hit the sidewalk and with what acceleration, thanks to Galileo and Newton. Yet one must thank the TV meteorologist, the computer simulations and any time series analysis methods the meteorologist might have used, on which to base the weather forecast.

In mathematical information theory "entropy" (we shall get to the use of this word in classical thermodynamics later) computes the amount of information measured in bits that one obtains from the probable outcome of an event. We can quantify this idea. Let p_i be some real number with values always between zero and one for each index $i = 1, 2, 3 \cdots$. What is the entropy we obtain after the toss of a fair coin? The formula to use is

$$H = -\sum_{i=1}^{n} p_i \log p_i, \qquad (6.13)$$

where n is some integer and logarithms usually are taken to base two. The value H is a kind of weighted average. From this we get for the toss of a fair coin

$$
\begin{aligned}
H &= -\sum_{i=1}^{2} p_i \log p_i & (6.14) \\
&= -\sum_{i=1}^{2} \frac{1}{2} \log \frac{1}{2} \\
&= -\frac{1}{2} \log \frac{1}{2} - \frac{1}{2} \log \frac{1}{2} \\
&= -(2)\frac{1}{2} \log \frac{1}{2} = -\log \frac{1}{2} & (6.15) \\
&= \log 2 = 1, & (6.16)
\end{aligned}
$$

or one bit of information, since we are taking logarithms to base two. Compare this to the amount of information we get from tossing a fair die:

$$H = -\sum_{i=1}^{6} p_i \log p_i \tag{6.17}$$

$$= -\sum_{i=1}^{6} \frac{1}{6} \log \frac{1}{6}$$

$$= -(6)\frac{1}{6} \log \frac{1}{6} = -\log \frac{1}{6}$$

$$= \log 6 = 2.5849\ldots, \tag{6.18}$$

or about 2.585 bits by rounding up to the nearest thousandth, where the logarithm is taken base two. So the amount of information we derive from tossing a fair die exceeds the amount of information we get from tossing a fair coin. You can try an experiment on your own. Suppose there are twenty small strips of paper in a hat where each strip has a single whole number from one to twenty written on it. Each event, picking some strip with a single number on it from one to twenty, is equally likely to happen. How much information does one obtain from picking a number on a single strip of paper from the hat? Is it more or less than the entropy for the toss of a fair die?

Entropy as defined in classical thermodynamics and statistical mechanics differs somewhat in description from the description above that we are giving in terms of bits. The German physicist Rudolf Clausius (1822-1888) was among the first to describe entropy in thermodynamics

$$S = \frac{\delta Q}{T}, \tag{6.19}$$

where T is temperature and δQ is a change in heat, as being "waste heat" given off from work (he based many of his ideas on the Carnot engine). Clausius's result supports the Second Law of Thermodynamics, so that it can be stated as follows:

A thermodynamic process evolves from one equilibrium state to another equilibrium state only so as to cause an increase of entropy.

Later Ludwig Boltzmann (1844-1906) more or less grounded the foundations of this law in classical statistical mechanics. He showed through his mathematical derivations that entropy is a measure that depends upon a sum of probabilities for the microstates of particles,

$$S = -k \sum_{i} p_i \log p_i, \tag{6.20}$$

where k is Boltzmann's constant and where the microstates are determined by the classical momentum coordinates for the particles and where these momentum coordinates help to constitute a "phase space." In Boltzmann's description the classical particles described by the phase space follow what is called the Maxwell-Boltzmann probability distribution. This equation is very similar to the previous discrete entropy equation we considered:

$$H = -\sum_{i=1}^{n} p_i \log p_i. \tag{6.21}$$

In fact S seems to differ from H by a constant. This is no accident. In fact the physicist Rolf Landauer (1927-1999) was able to show that any attempt to erase bits of information from a computer in an irreversible way leads to a certain increase in thermodynamic entropy. The idea is called *Landauer's Principle*:

> *The erasure of one bit of information from a computer requires a quantity of energy that is bounded below by kT ln 2.*

Here the logarithm is to the base e (not to the base 2) and temperature T is in Kelvins.

Now getting back to Shannon's arguments, what about the mathematical entropy (that is, the entropy determined in bits of information) one obtains by transmitting a message of English text, say? This depends upon the letter frequencies in each message. In the English language English letters have differing frequencies and some letter frequencies usually will be higher in written text than other letter frequencies. For instance if you still read newspaper articles (I do, because I still am old fashioned that way) or if you read an online blog, you will find the letters "E" and "A" have a higher frequency of occurrence in the text you are reading than do "X" or "Z." The writer Edgar Allan Poe made use of this fact in his famous story "The Gold Bug" which contained a simple substitution cipher to hide English plaintext, while in "The Adventure of the Dancing Men," the brilliant sleuth Sherlock Holmes also confronts a simple substitution cipher composed of two dimensional dancing men in a variety of poses. Deciphering such encrypted messages depends strongly upon knowing the probable frequencies of appearance for each letter in any English text, where these values, usually between zero and one can be treated as probabilities where all these probabilities sum to one.

Shannon used this fact to come up with another form of entropy which today is called Shannon entropy, which answers the question: Given the frequencies for each letter, how many bits of information are required to encode a text string written in the English language? Here we shall give an example of the kinds of text strings to which Shannon's result relates (See Chapter V in the book *An Introduction to Information Theory: Symbols, Signals and Noise*, by John R. Pierce, cited in the Bibliography). Suppose

A B A B A B A B A B A B A B A B ...

happens to be an endless, repetitive string where the alternation between the two characters A and B never varies in time and where the probability for each character is $\frac{1}{2}$. We can shift one character to the left to get

B A B A B A B A B A B A B A B A ...

or shift five units to the right to get

A B A B A B A B A B A B A B A B ...

Let us illustrate what one means by ergodicity. Suppose you get together one hundred friends and relatives or more and that each of these individuals has his or her own penny for the experiment. At some set time each one of you begins to toss your own penny one hundred times during the course of a day or week, that is, so that we have times $t = 1, 2, 3, \ldots, 100$. For each coin toss out of the one hundred separate coin tosses each one of you records whether your own coin landed heads or tails. Let $x_1[1]$ be the random variable for your first coin toss[3], $x_1[2]$ for the second, $x_1[3]$ for your third, and so on all

---

[3]Each value must be either H for "Heads" or T for "Tails."

the way to $x_1[100]$, which is your last coin toss at time $t = 100$. Similarly for the second experimenter the expressions are

$$x_2[1], x_2[2], x_2[3], \cdots x_2[100]. \tag{6.22}$$

When we include all one hundred participants including you, we get

$$x_1[1], x_1[2], x_1[3], \cdots x_1[100], \tag{6.23}$$
$$x_2[1], x_2[2], x_2[3], \cdots x_2[100],$$
$$x_3[1], x_3[2], x_3[3], \cdots x_3[100],$$
$$\cdots\cdots\cdots$$
$$x_{100}[1], x_{100}[2], x_{100}[3], \cdots x_{100}[100]. \tag{6.24}$$

At each time $t = t_0 \in [1, 100]$ each of you writes down in a table whether they got Heads or Tails for that particular toss and each of you will have one hundred of these results for your experiment. Now all of you sit down to calculate what the average number of heads is all of you got for any such $t = t_0$, say between times one and one hundred. So all of you calculate how many number of times you got heads for

$$x_1[t_0], x_2[t_0], x_3[t_0], \cdots x_{100}[t_0]. \tag{6.25}$$

You should find that the average number of Heads all of you got for $t = t_0$ is around fifty. One calls this an ensemble average.

Now each one of you finds separately from each other the average number of Heads you got for each of your experiments. For you that means you find out how many times you got heads within

$$x_1[1], x_1[2], x_1[3], \cdots x_1[100]. \tag{6.26}$$

You should find that for your part of the overall experiment, Heads showed up close to fifty times. Since the actual probabilities in both cases are close to 0.5, the random sequence can be shown to be ergodic as the number of trials goes to infinity ($t \to \infty$). The experimental results get more accurate when you increase the number of trials from one hundred to one thousand, to ten thousand and to ten million.

We can generalize this idea. Let each of the sequences

$$x_1[1], x_1[2], x_1[3], \cdots \tag{6.27}$$
$$x_2[1], x_2[2], x_2[3], \cdots$$
$$x_3[1], x_3[2], x_3[3], \cdots$$
$$\cdots\cdots\cdots$$
$$x_j[1], x_j[2], x_j[3], \cdots \tag{6.28}$$

be stationary random sequences for integer j, $1 \le j < \infty$, where one computes the time averages horizontally and the ensemble average along the vertical direction at any $t = t_0$. If these two averages are equal we have ergodicity.

One should note though that in statistical signal processing there are several different kinds of ergodicity, such as ergodicity in the mean, ergodicity in distribution, ergodicity in correlation, etc.

Ergodicity is an extremely important idea in statistical signal processing and in time series analysis and Shannon knew it, just as he knew the true importance of statistical correlation, that the statement "Correlation is causation" is false, when one deals with several random variables that might be dependent. Why? Because statisticians might find correlations between the rise in a stork population and the rise of human infant births in the same country, but that does not mean that the increase in the stork population had something to do with the rise in the number of human births.

Yet billions of humans today around the world formulate their opinions and very rigid biases on almost everything, politics, religion, violent crime, human intelligence, race or social status, goods and services, etc., based on society's powerful opinion makers who frequently use different averages and statistical correlations to distort the meaning or interpretation of statistical data. Marketers can use misinterpretation of data to make people who are prone already to biases to believe things that do not have to be true. For example suppose some data analyst is hired by a little league coach to take batting averages for children trying out for the team, then interpret results from the data. If the analyst finds that the batting averages for most of the children under seventeen years of age are lower than those for most of the children who are between seventeen and nineteen years of age, the data analyst with the right kind of data might be able to convince the coach the younger children did not hit the ball as far "on average," because they are all smaller kids and less physically developed, when the truth might be that *all* the younger children in the analysis were near sighted and needed glasses.

Potential customers might be told that eighty six percent of some consumers who were polled favored a certain product over another, when in fact the poll might not be from a truly random sample of the population. Millions of people who access various media such as RSS feeds, blogs or television and radio news broadcasts, will read or hear in endless repetition decade after decade, how members of some racial, ethnic, religious or social group, Group X, say, has a higher percentage of violent crime than do members of some other Group Y. But whereas statistics might show this is true overall "on average," such extrapolations are false for millions of individual cases within Group X, when the behaviors of particular individuals in Group X are examined over a long span of time. That is, if we *do not* have ergodicity then there will be always fixed times t_0 where this is large, such that the collection

$$x_1[t_0], x_2[t_0], \cdots x_j[t_0], \cdots \tag{6.29}$$

where j is large, will not exhibit the same statistical behavior as does

$$x_1[1], x_1[2], x_1[3], \cdots \tag{6.30}$$
$$x_2[1], x_2[2], x_2[3], \cdots$$
$$x_3[1], x_3[2], x_3[3], \cdots$$
$$\cdots\cdots\cdots$$
$$x_j[1], x_j[2], x_j[3], \cdots . \tag{6.31}$$

As we have seen ergodicity holds when the ensemble average is the same as the averaging over time. So one can select any particular ensemble to see that it shares the same statistical property as that of the other ensembles. So when ensembles *fail* to be ergodic, one cannot make valid assumptions that all the individual members of one particular ensemble will exhibit statistical properties exactly like those of all the other members in all the other ensembles over time.

Shannon entropy then is a means to calculate the average least number of information bits required to encode a text string, where one finds this value by using the frequencies of the letters.

One easily can code for a Shannon entropy calculator, perhaps programming the algorithm in MATLAB, Octave or in C, to output the Shannon entropy in bits or in bytes, for example for the string

`WAR BETWEEN NORTH AND SOUTH`

we already have considered. It turns out to be a minimum of 3.56 bits with rounding up to the nearest hundredth, which is less than one half byte. In comparison to store this same string WAR BETWEEN NORTH AND SOUTH in an MS Notepad text file takes up about 4×2^{10} bytes of disk space.

Using considerations such as these and other mathematical methods Shannon came up with significant results first for discrete signals such as messages sent by telegraph or teletype, that relate the speed of a message through a communication channel to the capacity of the channel. In his paper (See the Bibliography) he defines a communication channel as being a system through which one sends symbols such as integers or letters in such a way that each symbol in the message such as dots or dashes, is selected and has some discrete time duration measured in seconds. Shannon then defined the capacity of the channel to be the maximum rate at which the channel can transmit information. Mathematically he defined it as being

$$C = \lim_{T \to \infty} \frac{\log N(T)}{T}. \tag{6.32}$$

The value $N(T)$ is just the number of permissible signals that are T seconds in duration. Shannon then goes on to calculate tables for digrams, trigrams and even for whole words (See the book by Helen Gaines, in the Bibliography), estimating the corresponding frequencies of occurrence by using transition probabilities for Markov chains, which in this case describes the behavior of the stochastic or random sequences that are involved. Actually it is known that, when the transition probabilities converge to a definite state after so many successive stages of transition, the Markov chain is an ergodic Markov chain. Shannon precisely was considering the behavior of ergodic message sources in his paper.

Then after he discusses the role of entropy and also "conditional" entropy, which is for two or more random variables that are dependent (See the Appendix Section), Shannon explores with mathematics the entropy of various ergodic information sources and then the encoding and decoding of English text.

After this mathematical treatment Shannon arrives at his "Fundamental Theorem for a Noiseless Channel (See Bibliography), which he also proves:

> (Rephrased) *Let a message source have an entropy of H bits per symbol and a channel capacity of C bits per second. Then the output for the message source can be encoded in such a way so as to transmit $\frac{C}{H} - \epsilon$ symbols per second across a noiseless channel, where $\epsilon > 0$ can be taken arbitrarily as small as we please, and where transmission cannot occur for an average rate that exceeds $\frac{C}{H}$.*

A second important result was Shannon's treatment of discrete ergodic message sources where the signals are sent encoded through a noisy channel. In a second theorem he proved that when $H \leq C$ there exists an encoding for which the noise will be arbitrarily small, while if $H > C$ there exists an encoding method for which the noise will be bounded above by

$$H - C + \epsilon, \tag{6.33}$$

where $\epsilon > 0$ can be an arbitrarily small real number, where this is the best one can do under the circumstances, since any noise reduction bounded above by $H - C$ instead of by that in Eqtn. () is not possible.

Shannon's results in electronics communications were just as fruitful intellectually as was the invention of the transistor. Thanks to his results there has been a plethora of technological applications for landline phones, faxes, digital media such as music CDs and movies stored on DVD, cellular telephones, Wifi, Internet data packet transmission and digital TV and High Definition television. Then again we mention the impact on space probes, such as Voyager 1 and Voyager 2. The relationship between the entropy of an ergodic message source and the channel capacity has become as vital a concept in telecommunications as has the relationship between inertial mass, velocity and the speed of light in physics. Moreover each time some human being uses a smart phone to web surf and downloads a web page with noise free ease and with the aid of some remote web server, or listens to a music concert on a Bluetooth device that has a clear, noise free audio signal, he or she is part of a cybernetic system.

The significance of Einstein's discoveries are all around us in the observable physical universe of stars, planets, quasars and black holes, as gravitational lensing, gravitational waves and black holes all have been confirmed to exist by astronomers and astrophysicists who have worked with diligence at various observatories around the world and who have used everything from telescopes to laser interferometers such as the one at LIGO (Laser Interferometer Gravitational Observatory), to detect any deep space phenomena related to the predictions made by the General Theory of Relativity. On the other hand in the realms of electrical engineering, data communications and information technology, Shannon's theorems can help to determine the sound quality of an audio violin recording by violinist Joshua Bell, or the quality of a visual image of the dwarf planet Ceres. Shannon's results were profound, whether we really appreciate its technological implications or if we could care less about that but just to sit in one's entertainment room to enjoy an audio recording of Brahmss Violin Concerto in D Major, not caring at all about the difference between entropy and a logarithm.

One other paper of note by Shannon in 1949 was "Communication Theory of Secrecy Systems." This paper had an illuminating impact on the reliability of simple substitution ciphers to keep a message secret. As it turns out the longer one uses the same simple substitution encryption method and the same encryption key over time, the easier it will be for an opponent to decipher the plaintext for the actual secret message. Shannon proposed the use of "one time key pads," where the encryption key actually comes close to having the same length the message has and so that the encryption key is used only once. One algorithm through which one can accomplish this is called an XOR encryption algorithm.

6.4 Examples of Feedback and Control Systems

When it comes to cybernetics the Internet of Things has engendered a kind of synthesis between or coalescence of electronics, digital media, business and healthcare, computers and computer networks, automated systems, transportation systems, personal security and cyber security. One checks the Dow Jones or Nasdaq figures on a smart phone or laptop toward the end of the business day, sharing data communication between the phone and a remote web server. Businesses and banks change stock portfolios and the status of bank accounts, as responses to stimuli given as input from the investor. A patient sends an email to the doctor who is boarding a plane for a cross country trip to a medical conference. A parent at work uses a laptop to check the security status of the home apartment, perhaps with Bluetooth or remote sensing technology. Is someone burglarizing the place? Shoot off a text message to a security company or call the police on a cell phone. Is a rude neighbor snooping around in the back yard? Use an application to turn on your remote recording of a virtual, barking guard dog installed above near the kitchen window. What if some parent's child is taking too long to get home from school? A concerned mother can check the child's active smart phone remotely to ensure the child is not engaging in any unwise chats online with creepy strangers or to determine the whereabouts of little Billie or Angie with remote GPS technology,

High tech prophets predict that soon the elderly will be able to have remote, virtual physical checkups from their doctor without leaving the bedroom. It is possible even for a "virtual human nurse" or receptionist, both as embodied conversational agents, to make hospital appointments remotely for bedridden patients should the need arise. All this at present and in future comprise or will comprise cybernetic systems. In all these scenarios Shannon's theorems and the ideas of Norbert Wiener weave through to connect them all as with a common thread. A cyborg is not just The Terminator on DVD. It also is a human mom or a dad using a remote GPS or Bluetooth device, or an investor connecting by smart phone to a remote server, or even a JPL systems control engineer receiving remote data or telemetry from a space probe. Wiener's ideas on cybernetics are a part because all these diverse systems are connected via impulse and response mechanisms of one kind or another, along with feedback. But Shannon's ideas figure as well because with WiFi, broadband or wireless transmission there are noisy or noiseless channels through which one sends encoded messages.

In fact there are many different kinds of cybernetic systems, and all of them do not depend necessarily upon Internet communication, as we now shall see.

6.4.1 Thermostats

A human or even an animal that interacts in the manner of feedback and control with a thermostat is one of the simplest kinds of cybernetic systems. Thermostats have been in use since the very early part of the seventeenth century, sometimes to regulate the temperature in poultry incubators. Thermostats use different approaches. The earliest ones used mercury levels in a tube. Centuries later two-wire thermostats had been invented along with thermostats that operated by measuring the temperature changes of metal strips. Today digital thermostats are commonplace inside many modern homes. One of the earliest home thermostats I recall as a child growing up in Massachusetts in the nineteen fifties was the Honeywell T87. a neat little device with a round shape that always reminded me of an alien spaceship.

6.4.2 Electronic Alarms

In the sixties one could watch a TV crime drama in which some inept burglars would sneak through a window and in so doing, activate an alarm. During this time there also were movies like *Topaki*, about diamond thieves who were ingenious at avoiding the tripping of an alarm wire or photoelectric sensor and alarm system. Security and fire alarms have become more sophisticated over the years. Yet these even comprise feedback and control systems of a sort.

6.4.3   Is Your PC getting too Hot?

On most laptops there is a feedback and control mechanism that turns off the machine if it overheats on a hot day. This even has happened to me on occasion with legacy Windows XP laptops.

6.4.4   Cruise Control

Who has taken a long interstate trip inside an automobile without using cruise control on the highway? At times it might help to prevent one from getting a speeding ticket when the speed is set properly.

6.4.5   Positive Train Control

Positive train control (PTC) is a feedback and control system that would prevent the possibility of fatal train crashes or collisions. Various data values for train speed, navigation, track conditions, terrain, etc., would be updated regularly to enable the train to learn in a sense about possible errors in order to make any necessary corrections. One could describe such a train along with the passengers, the human engineer and even any onboard computer, as being a cybernetic system.

6.5   Cybernetics and Decision Support Systems

Various enterprises today, online large retail chains, pharmaceutical and bioengineering companies in the healthcare company, universities, even governments, can amass huge volumes of important data that can amount to gigabytes, terabytes, even petabytes of information that needs to be analyzed. This is in very stark contrast to enterprises that focus mainly on transactional data. For instance when you go online to your favorite online retailer, you might buy a hat or a book novel. The remote web server sends your browser the appropriate web pages to manage everything, your purchasing choices, your login sessions, your credit card verification, etc. Then all this data information is stored on a remote backend database server where any data updates are managed by client to server database queries. One calls this Online Transactional Processing.

In comparison decision support systems is what one means more or less by a relational database management system that is dedicated to the task of storing data in a form that makes it accessible to statistical analysis for the purpose of human decision making, in contrast to merely processing online data transactions (where this hopefully is *secure*) from moment to moment. Usually the decision support system stores the data in what one calls data warehouses, data stores or data marts. This kind of processing of data is called Online Analytical Processing (OLAP). Once the statistical analysis and predictive modeling or forecasting is performed, policymakers who might be corporate executives or department managers, financial officers or government officials, can make decisions based upon what knowledge is discovered from the stored data. One example of this is the massive amounts of data stored that retains information about terrorist activities since 9-11. Massive amounts of data stored in such database systems can be subjected to statistical analysis in order to predict or to estimate future terrorist behavior.

How does cybernetics relate to decision support systems?

Is a relational database a robot? Certainly not. Is it a cyborg? Perhaps not all by itself, but all human computer interactions or human to computer interfaces, comprise a cyborg of sorts. Every time a database administrator updates some inventory record in a database and anytime a data scientist or data analyst runs some data from the database through some artificial neural network or data classification algorithm, he or she along with the database system have become cyborgs.

6.5.1 Cybernetics and Operations Research

Operations research (which sometimes is called management science), is a branch of mathematics that enables applied mathematicians in this area to solve various complex organizational problems of high relevance to those who have to make serious organizational decisions. These people could be upper level management in the energy industry, or managers within agricultural corporations that have enormous farms to manage or even military admirals who must deploy sailors, naval vessels and war materials to a distant ocean in minimal time.

1. What is the best way some huge plot of arable land ought to be allocated, and how should various crops be grown on it, if the land and all these different crops have cost constraints?

2. How can soldiers, material and equipment be deployed best in minimal time during a war?

3. What can be done to minimize the wait times for some large bank branch's morning customers?

4. A company wants a new branch office in a different city. The executives want to arrange as much furniture and office machinery as possible into the new building but with the least amount of office space allocation. What is the least costly way they can do this?

5. What is the best way to schedule or to route plane flights and interstate trucking services, so as to minimize time and fuel utilization?

Problems like these usually are treated with mathematical techniques found in linear programming, Markov chain and martingale methods, queueing theory and the related discipline known as the mathematical theory of games. Four of the earliest pioneers in game theory and as researchers and authors, were John von Neumann, author of the monumental tome *Mathematical Foundations of Quantum Mechanics*[4], inventor of the Eniac and a consultant to the Rand Corporation, Oskar Morgenstern, John Nash who wrote a seminal paper in this field and Samuel Karlin, author of the two volume tome *Mathematical Methods and Theory of Games, Programming and Economics.*

---

[4]Originally published in 1932 and in German as *Mathematische Grundlagen der Quantenmechanik*

With mathematical models people in operations research find solutions to problems like these in business and government, to help managers, policymakers and others to make decisions. Yet at least one electrical engineer and applied mathematician from MIT, Leonard Kleinrock, found applications for queueing theory in computer networks. He also wrote seminal papers on the application of queueing theory to computer networks back in the nineteen sixties (See Bibliography).

The thing is that something like an airline company along with all its baggage clerks, pilots, engineers, schedulers, maintenance personnel, computer systems, etc., comprise a cybernetic system, and operations research describes exactly how to optimize various aspects of that system, aspects such as fuel, allocation space, customer interarrival and wait times at airports, inventory costs, etc.

In this chapter we have seen how complex and seemingly disparate systems, such as technological, business, government or electromechanical systems and humans, especially when all these serve together as the different components of some larger system of interactions, can be described as being a cybernetic system as this was conceived by Norbert Wiener. We also have seen that some cybernetic systems depend upon the transmission and reception of encoded data and how Claude Shannon established the mathematical means as to how this can be done across both noisy and noiseless channels of communication. Both these new disciplines however, information theory and cybernetics, germinated at a time when digital computers that process digital data was in an initial stage. As we have seen, Vannevar Bush's Differential Analyzer in fact was a sophisticated analog computer not a digital computer.

Many years ago mechanized robots of various kinds also were analog machines, some of which truly were remarkable in their learning ability, such as Claude Shannon's wheeled analog robot Theseus, and varied in their possibilities. We discuss them in the next chapter.

Chapter 7

Cybernetics and Analog Robots

This Chapter is short because the robots we discuss here are no longer very active areas of research as once was the case back in the nineteen forties and fifties.

Analog robots were built from very simple design concepts that exclude the use of CPUs, processors, operating systems, actuators and computer code. Actually robots of this type frequently were built only with a minimum of moving parts: Vacuum tubes and photoelectric cells and sensors or small microphones to detect sounds, batteries, wheels and chassis. Analog robots are based on simple electromechanical concepts so that the device responds in sometimes unpredictable ways to an external stimulus, such as reacting to a light or a sudden noise.

7.1   Theseus, the Electromechanical Mouse

Claude Shannon not only was a formidable applied mathematician but also someone who had a flair and a talent for physical invention. One of his inventions was a mechanical device that exhibited a surprising but apparently useless cycle of impulse and response. It was a little box that, when someone activated it (impulse), a small mechanical hand would pop out the box to turn off the device (response). No doubt a mechanical device like this must have elicited intense discussion and speculation within Shannon's professional and social circles, as indeed a Rube Goldberg machine might have done, although these latter hypothetical machines did perform some useful but very simple function, albeit by means of ponderously complex mechanisms.

But Shannon also invented a far more useful and ingenious contraption, a little analog, electromechanical robot mouse named Theseus. In Greek mythology Theseus like Heracles had many adventures, one of which was the slaying of the Minotaur and then winning the heart of Princess Ariadne, daughter of King Minos of Crete. Unfortunately for any romantics their love story did not lead to their living "Happily ever after." The German composer Richard Strauss in his comic opera *Ariadne auf Naxos*, treats events in Ariadne's life after Thesues abandons her on Naxos.

Like its predecessor from Greek mythology, the Theseus analog robot was able to navigate its way through a complex square maze, although not by a long thread given to it! This analog robot displayed a distinct ability to mimic a kind of learning process, that depended upon what relays and switches were activated.

It is fascinating how this early analog robot was able to find its way through a maze without having the sophistication of the digitized robots we today see in use. How exactly did Theseus remember how to navigate its way to its target through paths that had been modified? Did its relays and switches follow some specific sequence of activation? It makes one wonder if in some primitive way it was able either deliberately or by accident, to mimic or to emulate some neuron and synapse based, electrochemical circuits within the brain of some real mouse, so that these real animals could perform in a similar manner.

7.2   William Grey Walter and his Analog Robots Elsie and Cora

Today computer science, robotics and neuroscience are three separate disciplines that frequently have ideas that merge, given a certain problem. When it comes to this type of approach to research William Grey Walter (1910-1977) was a cyberneticist and a neuroscientist who was ahead of his time. Proficient in electromechanical ingenuity he modified and extended the capability of the electroencephalograph. One benefit of his innovation was that his improvements on the EEG helped to map the brain regions responsible for the convulsions suffered by epileptic patients.

William Grey Walter (1910–1977, a man who in some photographs bore a very strong resemblance to Leon Trotsky) was born an American citizen in Kansas, MO., but he was educated in the UK. It was there that he did most of his work on brain waves, which inspired him actually to start building electromechanical robots, since he believed that by studying the various stimulus and response or feedback and control mechanisms of these machines he would understand the human brain much better.

Walter's first wheeled robot, was a mechanical tortoise named Elsie he developed along with other similar robots in the UK in the late forties and early fifties, around the same time period that the applied mathematics and electrical engineering communities in the States were peering into Claude Shannon's papers on information theory. This autonomous, analog machine sported vacuum tubes, photoelectric sensors and tactile antennae as the contemporary state of the art in its technology. It also was a cybernetic system in that its frequently unpredictable, physical responses were stimulated by various light sources. Walter then trained a similar analog robot (named Cora) that he built to exhibit conditioned response behaviors by making erratic motions and uttering audible whistles. The intriguing thing about Walter's autonomous devices was that they mimicked very well the learning behaviors or conditioned responses of primitive animals, such as the dogs used in the experiments of the Russian scientist Ivan Pavlov. Like Shannon's electromechanical robot Theseus that was capable of navigating through a maze, Walter's robots seemed to exhibit the ability to learn.

7.3   Elektro the Smoking Robot

Today a child has his or her dog with which to play. But at the New York World's Fair in 1940 Elektro had Sparko (See the Wiki link in the Preface). Back then Elektro was an electromechanical humanoid robot seven feet in height and Sparko was his own version of an automated electronic dog. Built by engineers at Westinghouse Electric Corporation, Elektro was a robot that weighed somewhere between two hundred and three hundred pounds, and was voice activated. He had photoelectric eyes that helped it to distinguish between two different colors.

Elektro also walked and talked. The automaton could move its mouth as it spoke but the vocabulary and subject matter was restricted to a 78 RPM record played on a record player. The robot actually could puff away on a cigarette and he could move his head and arms, something that might have caused some consternation if a human visitor got too close to the automaton. By his own spoken admission to the viewing audience Elektro had in him forty-eight electrical relays for a "brain," arranged into some sort of old fashioned PBX kind of switching circuit. This is an example of a robot with a "brain" that behaved more or less like a finite state machine.

Elektro was an impulse and response mechanism. The spoken commands of the human operator served as the input and the specific output depended upon what was input by voice command into the machine in a kind of voice activated remote control system.

A similar robot was the imposing Alpha the Robot, which was a technological sensation at Macy's department store in Manhattan New York back in 1934. Built by British engineer and inventor Harry May, Alpha the Robot was able to stand and sit, talk in a very intimidating voice and fire off a pistol.

Today Elektro the Smoking Robot is at a museum in Mansfield, Ohio. But those who would like to see Elektro in action can find him on YouTube, at the time of this writing. The reader can find more information about Alpha the Robot at www.cyberneticzoo.com.

7.4 The "Hopkins Beast"

The analog feedback and control robots that Claude Shannon and William Grey Walter developed in the nineteen forties and fifties were replaced in the sixties with new switching circuits and logic gates that depended upon the transistor, not upon the slower technology of electromechanical switches and vacuum tubes.

In 1964–65 one such example of a transistor based autonomous machine that was developed at Johns Hopkins University was called the "Hopkins Beast." The prototype was a transistor based, wheeled robot that responded to sound or acoustic inputs within one of the corridors in the Applied Physics Laboratory. The stimulus and response mechanisms enabled it to navigate its way through the corridor until it found a specially adapted electrical wall outlet at which it would recharge. One might say it was a technical case of not having to lead a mechanical beast to water nor having to force it to drink. Subsequent versions invented at the Laboratory also used more elaborate transistor based logic gates along with photocells.

Can such primitive autonomous devices today be of any use? Probably, given the right circumstances, environment or setting. For over a decade schools and college campuses within the US such as at Sandy Hook in New York and at Virginia Tech, have become the backdrop for mass killings at the hands of psychopathic or crazed gunmen. It is conceivable that a device similar to the Hopkins Beast could be designed to monitor school hallways and corridors for unusual noises, such as rapid gunfire from some crazed gunman in the hallway, and so that a built in stimulus and response mechanism might enable the robot to send an immediate automated emergency email alert or phone text message warning to campus police or to department heads. Such a device also might be able to emit a piercing audio alarm to warn others in the immediate area who might be in classrooms.

7.5 The Stanford Cart

The Stanford Cart was a mobile robot developed at Stanford University's Artificial Intelligence Laboratory (SAIL) in the nineteen seventies. It was maneuvered remotely on four bicycle wheels and was able to fix its location relative to some distant trees in the vicinity of the parking lot. However it did not perform so well in terms of self-navigation when one measures its performance by the successes of the more primitive Boolean circuit and relay based "signal and response" cybernetic machines like Theseus, Elsie and the Hopkins Beast. Yet at Stanford the times were part of the early days of an AI research that utilized the earlier kinds of digital based technology and television cameras, which in comparison to the microprocessors and high definition digital television systems of today, were far more primitive and limited.

The Cart navigated by determining its position relative to some fixed external object, such as a distant landscape of trees or a white line in the middle of a road. Even though its early efforts seemed limited, subsequent studies provided researchers with vital information as to how to get such a machine to learn, to adapt and to recognize the different features in a physical terrain or landscape. Eventually such research at Stanford and elsewhere led to the development of driverless automobiles.

7.6 Shakey

The Cart, a wheeled robot about five feet in height was controlled remotely by computer. It used a television camera and antenna, wheels and a simple suspension system for indoor navigation within interior rooms that had walls, doors and physical blocks. Its "edge based" navigation and computer vision technology helped it to locate the physical blocks in the room. Certainly this is a far hue and cry from what wheeled robots can do with today's computer vision systems. One should keep in mind thought that these early efforts were only the very beginnings of autonomous vehicle technology.

7.7 Summary

In this Chapter we have seen how the research by scientists like Norbert Wiener and William Grey Walter helped them to glean valuable information on cybernetic systems and how they function, and on how the stimulus and response behaviors of electromechanical robots mimicked successfully the behavior of small animals and even possibly, brain wave or nerve signal activity at work in the human brain. We learned in Chapter 6 about NASA's spectacular successes with one autonomous deep space probe after another from Voyager 1 to the Mars Rover. We have seen also in Chapter 7 how the work of Claude Shannon formed the foundation for the boon in complex digital systems, communication and information technology that exploded into production in the decades after his two important papers.

But so far we have discussed for the most part, robots that have played roles either in the laboratory setting or in the exploration of deep space. But in the next three chapters we shall discuss the explosive growth of robotics within other human environments, such as in the military, in hospitals, in prosthetics and in industrial manufacturing.

Chapter 8

Autonomous Vehicles and Drones

8.1 Who is Driving?

Who is driving the car? Who is piloting the plane?

Today there are new autonomous vehicles in operation so that no human auto drivers and airplane pilots need to apply for employment positions to be a driver or pilot.

After the many heralded successes that NASA engineers and their project managers enjoyed from 1978 to 2015 with their unmanned space probes from the Voyager to the Project Rosetta missions, it comes as no surprise that low cost, robotics control systems software and hardware would find its way to the free market and into the hands of robot enthusiasts, undergraduates in college mechanical engineering and electrical engineering departments and at IT startups. There have been precedents. The Internet for instance began as a Defense Department computer communications initiative in the days of the Cold War. But soon it evolved into the ARPANET and from that into eventual public Internet use for the development of the World Wide Web along with its protocols like TCP/IP.

Combined with sophisticated remote sensors, robotic vision technology, GPS and robot navigation applications, more and more everyday people began to explore the unlimited possibilities of new robotics research and development along with AI. Many of these other more private and sometimes maverick research efforts though, were spurred on, incentivized by various government and sometimes private funding challenges and competitions, such as with DARPA and sometimes by the National Science Foundation. It is easy to succumb to the notion that if Vannevar Bush was alive today he might be pleased by this.

There were many competitions and developments in unmanned vehicles and wheeled robots from the late nineties until 2008, too many to cite them all. Nevertheless in this Chapter we mention briefly some of the more outstanding achievements in these areas.

8.2 Kratos, PAVE, The Subjugator and uBot-5

8.2.1 Kratos

In 2008 Kratos, a wheeled navigational robot built by undergraduate student members of the Princeton Autonomous Vehicle Engineering team, won one of the awards for collision and/or obstacle avoidance, in the Intelligent Ground Vehicle Competition. At first sight Kratos reminds one of the unmanned roving vehicles NASA sent to the Martian surface, like the Mars Rover. One difference though was that Kratos depended upon GPS technology for navigation.

8.2.2   Prospect 12

As we have seen in a previous chapter, mechanical engineers, computer scientists and artificial intelligence researchers have been working on the design and construction of autonomous vehicles since the early efforts were made with the Hopkins Beast at John Hopkins University and with Stanford University's CART. However with technological advancement of things like transistors, microprocessors, logic controllers and operating systems, autonomous vehicles have become far more sophisticated than the best efforts of the nineteen fifties and sixties. One example of this is Prospect 12, which was an autonomous vehicle research project conducted by Princeton students on the Princeton Autonomous Vehicle Engineering team in 2007. As software engineers Kyle Johnson and Trevor Taylor discuss in the book *Professional Microsoft Robotics Developer Studio*, the young participants in DARPA'a Urban Challenge retrofitted a Ford Hybrid with special cameras for vision and with other devices that were in communication with dual-core servers. At least some of the software development required Microsoft's Robotics Developer Studio's GUI applications, which however does allow developers to modify *C* code for various programming tasks.

8.2.3   The Subjugator

The Subjugator was a small autonomous submarine. Built at the Machine Intelligence Laboratory by students at the University of Florida, and depended on Microsoft's Robotics Developer Studio software for controller its development. Its operating system mainly was Windows XP, which by now of course has become a legacy application. Nevertheless the Subjugator had a highly sophisticated sensor system to detect height and internal temperature, among other things, as well as hydrophones and cameras for "eyes."

The Subjugator it would seem, has considerable potential for practical use. With further modifications it could be deployed to collect data for climate change research or for research in meteorology or in oceanography in academia at the very least at the undergraduate or graduate or level, as it could be used to detect temperature differences in the ocean from one distant location to another, or to monitor marine life or perhaps one could use it to monitor the activity of great white sharks along a coastline or at a beach where there is much human activity.

While The Subjugator was an aquatic version of an autonomous machine, the uBot-5 was a two-wheeled robot on terra firma, usually in an indoor environment. Developed by students at the Laboratory for Perceptual Robotics at the University of Massachusetts Amherst, this robot, developed as well by the same Microsoft application, was programmed with a servo controller that enabled it to move adroitly on its two wheels, performing a kind of balancing act like an inverted pendulum. The uBot-5 had two mechanical arms. For a head it used a kind of navigational monitor. This particular robot had potential for further development so that it could be used, perhaps, in the health care industry.

8.3 Self-Driving Cars

Google's claim to fame lies mostly in the fact that the company's software engineers came up with some of the most efficient information retrieval systems, text mining algorithms and search engines in the world. Even so in the mid 2000s and after that Google has expanded its research across the vast and variegated frontiers of artificial intelligence with its self-driving car called the Google SDC. Sebastian Thrun from the Stanford Artificial Intelligence Laboratory supervised the project but there were several other AI and robotics researchers who worked on the Google SDC.

In 2005 while still at Stanford Sebastian Thrun and his robotics team along with engineers at the Volkswagon company's VWERL laboratory, won in the DARPA Grand Challenge for their autonomous vehicle robot named Stanley. This driverless car was able to fix its location and to navigate in part by using digital maps along with a Lidar remote sensing system. Thrun is a pioneer in probabilistic robotics. Before the advent of probabilistic robotics many robots, in particular industrial robots for instance, were programmed to perform fixed tasks. Years ago conventional robot servomechanisms and actuators along with program code would help an industrial robot arm to fix its location in space, or to determine how many degrees of rotation it would have to swing to place a component part on a manufactured item such as a car. Yet robots like these earlier versions in manufacturing were not good when it came to reacting to uncertainty, such as if a tool it had been reaching for were to roll over suddenly to the floor due to a gust of wind when someone opened a nearby door.

Probabilistic robotics is a rather new applied mathematical discipline in robotics one uses to find ways to help a machine deal successfully with uncertainty or to use another word, *noise*. This in fact is what driverless cars in the near future must do if they are to function successfully and without frequent accidents or collisions on the highways of a city, town, province or state. Even now though one can envision scenarios in which driverless vehicles show great promise. Autonomous cars and buses could shuttle students and faculty members from building to building on a large university campus separated from public roads. These machines even could be used as private transportation services within a privately gated community. This technology even has the potential to put many a golf caddy out of work.

8.4 Unmanned Aerial Vehicles

After the terrorist bombing of the World Trade Center on September 1, 2001, the United States Air Force used Predator and Reaper drones with devastating lethality during the wars in Iraq and Afghanistan. Capable of both remote controlled operations and autonomous action, these drones sported a plethora of different kinds of hi-tech instruments for the purposes of surveillance, navigation and remote sensing: Temperature sensitive sensors, infrared and video cameras, laser rangefinders, GPS, and instruments for precision bombing and missile launching. The Air Force deployed hundreds of them and eventually the technology was shared by other US government agencies, including the Department of Homeland Security and US Customs and Border Patrol.

During the 2015–2016 US Presidential primary campaigns some politicians discussed quite heatedly their sentiments in which they expressed convictions that a wall ought to be build at the southern US border, to keep out illegal immigrants. It seems the Israelis have had considerable success in curtailing violent terrorist activity after the construction of their wall on the West Bank. On the other hand some have suggested that organizations like Hamas simply altered their tactics after the wall construction had begun.

Most of us have heard about Hadrian's Wall in Great Britain, the former Berlin Wall and about the Great Wall of China. Building a wall to keep out illegal immigrants might sound good as a campaign gimmick; nevertheless such an enormous building construction could be cost prohibitive if the US pays the bill for it, given the current gargantuan US federal deficit close to twenty trillion dollars. In fact building a wall, given that Mexico as a sovereign nation actually does refuse to pay for the monumental construction costs (a possibility some politicians fail to consider), also could prove to be just another serious federal budget buster.

One could consider as a less expensive alternative, however, the possibility of the use of autonomous and semiautonomous UAVs for continuous around the clock border surveillance, as well as a wireless surveillance sensor network, trustworthy US citizens serving as volunteers trained by qualified US Border Patrol officers, all to be deployed continuously along the border, to search day and night for illegal immigrants being smuggled across it.

Another potential use of remote controlled quadcopters is for the aerial tracking of various criminals such as bank robbers or armed robbers and burglars. One of the biggest problems for law enforcement is to try to catch a bank robber who committed the crime while heavily disguised. But just imagine the robber being tracked with stealth by a small aerial drone, disguised perhaps as a bird or an insect, as he or she makes the escape to an automobile, taxi or city bus. Ideally and if the miniature UAV is using some sort of remotely operated web based camera technology, the bank robber might be tracked directly to their home or to some other supposed "safe haven."

8.5   Domestic Quadcopter Drones

Today remote control quadcopters are so ubiquitous anyone with three or four hundred dollars can purchase one at Amazon, to add some cameras and, after configuring the software, launching it to take aerial photographs over some remote location. There has been talk within some online ecommerce companies of using quadcopters for the home or business delivery shipments of products ordered by a customer. Others in Africa actually have used remote controlled quadcopters to catch elephant and rhino poachers in the act as they engage in their cruel and illegal activities. Quadcopters can be a source of harmless enjoyment and utility for many, yet there have been serious issues raised by their use.

Although the widespread use of domestic drones such as quadcopters have raised safety and security concerns among many, still one can see the value in their use when they are properly in use. Some ecommerce and shipping companies large and small have begun to use drones to deliver goods and services, including the delivery of food and medicine. In contrast CyPhy (http://cyphyworks.com) is a new company that has developed "tethered" drones, meaning drones in the air suspended above ground but that are tethered to Earth by cables and that can remain suspended indefinitely at the same spot several hundred feet in the air. Security is enhanced through the enforcement of physical links for communication between the suspended drone and its Ground Control Station. Internet and network engineers can remember the days of the old wired Ethernet Although since it was wired technology it could not handle the larger bandwidth for today's wireless devices, still it would be impossible in most situations for a malicious hacker to use a wireless device to steal encryption keys or to hijack a session within a secure, wired LAN.

Can such a technology help to bring wireless technology to regions of the world that have little or no wireless, wifi or broadband technology? Perhaps time will tell.

8.6 Homeland Security and Cyber Security Issues

It is one thing for a robotics hobbyist to fly a quadcopter remotely to take pictures of sunset at a beach. Yet it is something else entirely for civilians to fly them near airports close to airplanes that are about to land or that are taking off, or too close to military installations, police stations or religious buildings. In a world as chaotic and as tempestuous as is this one, why take chances escalating a harmless situation of flying a quadcopter into a panic, angry misunderstanding or even into tragic situation? At the time of this writing some States have been considering regulating quadcopter use or having their human operators register their machine with a government agency.

Another serious consideration is the fact that all too frequently evil people employ both science and technology toward their own evil goals and ends. Airplanes have been used in the cause of smuggling contraband and for transporting illegal drugs across borders. Cybercriminals repeatedly use malware to compromise computer systems. Today still there looms the danger that terrorists might build a radioactive "dirty bomb." What is to stop some criminal or terrorist from using quadcopters toward a similar purpose? Already some have attempted to retrofit small arms weapons such as firearms to domestic drones that can be manipulated via remote control, according to the news media. Some nations even now have taken steps to deal with these dangers. For instance there are European countries that have trained eagles actually to attack autonomous drones that might pose a threat to life or some other danger. Yet all is not lost. Yes, perhaps terrorists and criminals can use technology for evil ends. There are others however, who can utilize the same tools to stop them.

Chapter 9

Robot Arms, Bipeds and Quadrupeds

9.1 Introduction

It is one thing for a movie screenwriter or even for a mathematician to conceive of robots that either can destroy the Earth as the robot Gort could do, or self-replicate themselves as they engage in space exploration as the theoretical space probes conceived by mathematician John von Neumann could do, but something else entirely for mechanical engineers and electrical engineers to design and to develop complex robots from an early conceptual stage to actual physical construction, robots that can be programmed to perform specific work or tasks. Yet even with this under consideration the first true working robots that were realized, designed and built in the United States, Europe and Japan after World War Two, were limited when it came to a robot machine's actual resemblance to humans. For one thing the early industrial robots, for example, only had a crude resemblance to one human body part, specifically the human arm.

9.2 Industrial Robots

The inventor and successful businessman George Devol (1912–2011) and the physicist and electrical engineer Joseph F. Engelberger (1925–2015) were the earliest visionary pioneers in the field of industrial robotics. When one contemplates their combined achievements in the exploration of the early robotics frontier within the nineteen fifties and sixties (research done largely within industrial and nonacademic corporate settings back in the fifties, one might add) one could consider them to be the Lewis and Clark of industrial robotics. Actually it was the inventor George Devol himself who developed the very first "Unimate" robot arm for industrial use. But together they forged ahead into the brand new territory of industrial robotics development with a sweeping success that led to the widespread acceptance of their new robotics technology across the United States, Europe and Japan.

These earliest robot arms developed by Devol and Engelberger for industrial use were large and massive, of the order of two tons. They were what engineers call "articulated" robots, meaning they were robot arms with actuators and that moved joints as a human might bend an elbow and rotate a wrist. The earliest versions performed simple programmed tasks, like picking up an automobile part from one location near the robot arm to translate, rotate or move the object to another nearby location (mechanical engineers working in robotics call this "pick and place"). These early robot arms also depended frequently upon "knowing" how to move in ways determined by angular degrees of rotation. The early programmers of industrial robots helped the robot to do this first by "teaching" it to move by particular predetermined angular degrees of rotation or by various predefined Cartesian coordinates. The total number of different translational or rotational motions that were allowed for the robot arm comprised the machine's "degrees of freedom." Clearly this expression *degrees of freedom* has a meaning that differs considerably from the same expression one finds in inferential statistics, such as when one finds the number of degrees of freedom one obtains for some parameter estimation of an expected value or mean, or in a test for goodness of fit. For the robot degrees of freedom means the number of translational and rotational directions in which the robot parts can move.

How though, did the robot arm "know" how to move? Through programming, although of a primitive kind, back in the nineteen fifties.

The memory storage devices for the earliest industrial robots of the nineteen fifties were magnetic drums that might have been no match for today's data intensive robotics, Big Data and AI systems. In fact the earlier magnetic drums could handle much less than one megabyte of memory per robot arm. Nevertheless these magnetic drums did suffice for the more simpler tasks that were required at the time, such as picking up and moving an automobile door handle from one location to another, or welding a door joint. A metal drum was coated with a thin layer of ferromagnetic material. As the drum rotated there were rows of electrically charged tape heads that emitted electrical pulses as they moved along tiny magnetized and polarized "particles" on the drum for the READ and WRITE operations.

Years after their initial breakthroughs in industrial robotics, George Devol and Joseph Engelberger witnessed the progress and continued growth and development of robots designed and built for applications in industry and manufacturing and especially in the automobile industry: Robot arms for pick and place, robot arms for spray painting, even for packing food products in boxes and storing consumer items on pallets for future shipment. Some of the companies that were inspired by and adopted robotic arms for factory use included General Motors in the US, where industrial robotic arms were used for spot welding of certain automobile parts, KUKA in West Germany and Hitachi in Japan. When this author was still a graduate student in computer science and information technology he had some small opportunity to observe how robot arms behave at least in computer simulations for one of my classes that covered computer simulation and animation (2D), although my student assignment at the time was involved more in creating animations and running numerical computer simulations specifically to develop various reliability models and other computer simulations for queueing systems and waiting lines (i. e., such as customers in banks, airports and palletizing robotic arms used in manufacturing plants and production centers) for industrial applications.

It goes without saying these early industrial robotic arm systems could function only inside specifically controlled physical environments. If the robot had to pick up a nut, a bolt or a hammer it had to know precisely where the object was each time and every time the robot reached for it. So just an open window that was opened suddenly by a human worker or a sudden wind gust from an open door could set things wrong. Or allow the object to be out of position by one millimeter or by one inch just once along some moving assembly line or conveyor belt and a human technician or quality control expert would have to be summoned. Today however we see robotic and AI systems in use such as "probabilistic robotics" and "deep learning" systems, or automatic control systems that can update information very rapidly for the purposes of adaptation and error correction, even in the presence of statistical noise or system error, as we shall see later on in this chapter.

9.3 Robots that Roll, Walk and Run

As we saw already in the previous section, industrial robotic arms were restricted in their movements and operated in very limited and controlled physical environments. These were and are *articulated robots*, heavy robotic arms connected firmly and either in a rigid position at the base or also so that the arm could rotate if necessary, but jointed elsewhere along its structure so that all these joints could move about freely. Sometimes there were three of these joints and for other robot arms there could be as many as six joints, for example three for translation by sliding in and out such as with a "prismatic joint" and three for rotation.

Back in the nineteen fifties and sixties these robotic arms did emulate to some extent the physical manner in which human arms and elbows do move. But today mechanical robotic systems have taken on a new generation of technological sophistication. Today there are robotic arms and hands that can perform surgery on human patients or that can hold a housefly carefully by the wing. Other robots have been constructed to climb stairs, to roll rubber balls across a table, to play ping pong, to welcome human guests to a business office, even to perform as automated pack mules, as does the quadruped robot Big Dog, which was developed at Boston Dynamics for the military. Doubtless these much newer, more sophisticated robots must have more than only three to six degrees of freedom. Correspondingly the algorithms by which they operate have become much more complex, so involved and intricate that those old magnetic drums never would suffice to handle memory. Here now we shall take a look at some of these robots, some of which have been designed to look and to function like humans in motion, but others that have been developed to take on the characteristics and motive features that many animals exhibit in the wild.

9.3.1 Polygonal Robots, Point Robots and Motion Planning

Whether a robot moves on mechanical wheels or on mechanical legs, there must be some generalized means and an algorithm by which one can design the robot motion behavior, so that the robot whether a mobile robot like a driverless car or a humanoid robot, can navigate successfully without resorting to undesired collisions into objects in a room or building or along a street or highway. Otherwise a robot that moves on legs or on wheels would have a really tough time trying to navigate its way from laboratory room to laboratory room, from street to street or from Martian crater to Martian crater. One frequently can do this by describing the robot motion at an abstract mathematical level, such as with abstract robots called *point robots* and convex *polygonal robots*. By "convex" we mean that a polygonal robot is an abstract robot that is represented by a convex polygon, where by "convex" we mean simply that if you connect any two points either inside or on the polygon with a line segment, that line segment must lie either entirely within and/or on the polygon or on the boundary of the polygon. To illustrate any circle, triangle, square, rhombus, quadrilateral with parallel sides, regular pentagon, hexagon, or trapezoid, etc., would be a convex polygon, while a two dimensional plane figure that looks like the block letters for "W," "E" or "Q" would not be a convex polygon. On the other hand a "point robot" derives from the coordinate system for the polygonal robot. On the two dimensional plane or in space the coordinate positions (x, y) (or (x, y, z) in the case of a convex polyhedron) for the polygonal vertices frequently are replaced with a point of origin such as $(0, 0)$ or $(0, 0, 0)$. If there is rotation also to consider as well as translation, one includes also an angle ϕ on the plane or three angles ϱ, θ, ϕ, in $3D$ space. The corresponding point robot coordinates on the plane or in space then become (x, y, ϕ) or $(x, y, z, \varrho, \theta, \phi)$ respectively. On the two dimensional plane then there can be three degrees of freedom while in three dimensional space there can be six degrees of freedom.

Robotics engineers then consider two spaces; one is called the *work space* (See Figure 9.1), the space in which the robot actually moves (i.e., without colliding into walls, furniture, tables, mailboxes, etc.), and the second space one calls the *configuration space* (See Figure 9.2), which is the space in which the coordinate parameters exist and are used to trace out the polygonal paths and in particular some optimal path, that connects the point of origin to the destination in a way that does not include any intersections into forbidden regions along the path[1]. These "forbidden regions" in the geometry would be areas in the configuration space at which the path passes through geometrical regions which are preserved for the forbidden obstacles such as tables, chairs and so on.

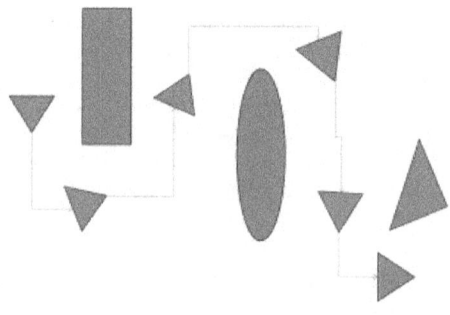

Figure 9.1: Example of a Work Space.

So add other fixed polygons within these two different spaces to represent things like fixed tables, chairs, parked cars, buildings, trees, etc., and one sees that one represents the point robot's successful navigational motion (by "successful" we mean so that the robot does not collide with a table or tree, for example) within a room or around the buildings, exterior objects and other physical structures on a college campus, as a polygonal curve that moves around all these other polygonal objects without crossing or intersecting any of these other polygons that represent tables, chairs, cars, trees, etc. There is an analogue to this in classical physics, where physicists distinguish the space in which physical objects move such as rockets, ships and billiard balls all given Cartesian coordinate positions

---

[1]In Geographic Information System databases a polygonal path sometimes is called a *linestring*.

(x, y, z) so one can tell where these physical things are located in physical space, and the other space in which the physicists study the position and momentum coordinates (x, y, z, p_x, p_y, p_z) for these physical objects. The best polygonal path in the configuration space will be some sequence of connected points on $\mathbb{R}^2$ for instance,

$$(0, 0, \phi_1), (x_1, y_1, \phi_2), (x_2, y_2, \phi_3), \ldots, (x_j, y_j, \phi_j), \tag{9.1}$$

$$0 \leq \phi_1, \phi_2, \phi_3, \ldots, \phi_j \leq 2\pi, \tag{9.2}$$

that will lead from the origin to the destination without passing through some forbidden region.

The branch of applied mathematics that treats the topics we cover here is called "dis-

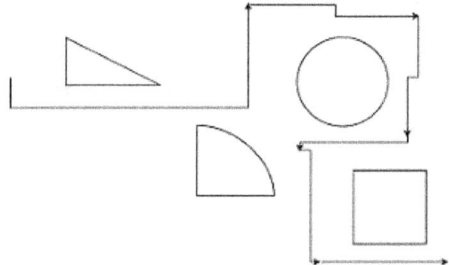

Figure 9.2: Example of a Configuration Space.

crete geometry" or "combinatorial computational geometry," and the kinds of coordinate transformations that are covered are from what mathematicians call an *affine* geometry[2]. Frequently the configuration spaces in question are denoted as

$$\mathbb{R}^2 \times SO(2), \tag{9.3}$$

and by

$$\mathbb{R}^3 \times SO(3), \tag{9.4}$$

for motion planning problems that have three and six degrees of freedom, respectively and where the $\times$ symbol is for the Cartesian product between two or more nonempty sets. For example with two sets A and B where

$$A = \{1, 2, 3\}, B = \{a, b\}, \tag{9.5}$$

we have the Cartesian product of ordered pairs

$$A \times B = \{(1, a), (1, b), (2, a), (2, b), (3, a), (3, b)\}. \tag{9.6}$$

For the natural numbers $\mathbb{N}$ we get a countably infinite set of ordered integer pairs

$$\mathbb{N} \times \mathbb{N} = \{(1, 1), (1, 2), (1, 3), \ldots, (2, 1), (2, 2), \ldots\}. \tag{9.7}$$

---

[2]There is another field of computational geometry used widely in the areas of computer graphics and numerical computing/fluid dynamics, which we do not treat here.

So $\mathbb{R}^2$, which just is another way of writing

$$\mathbb{R} \times \mathbb{R}, \tag{9.8}$$

denotes the set of all real coordinate points on the plane. Similarly $\mathbb{R}^3$, another way to write

$$\mathbb{R} \times \mathbb{R} \times \mathbb{R}, \tag{9.9}$$

denotes all the real three dimensional coordinate points in Euclidean space. The expression $SO(2)$ denotes the *special orthogonal group* of rotations of a vector by matrices on the plane and $SO(3)$ is the special orthogonal group of matrix rotations of vectors in three dimensional space. They describe rotations for coordinate transformations

$$(x, y) \rightarrow (x', y'), \tag{9.10}$$

$$(x, y, z) \rightarrow (x', y', z'), \tag{9.11}$$

that preserve the distances

$$d(x, y) \quad = \quad \sqrt{x^2 + y^2} = \sqrt{(x')^2 + (y')^2} \tag{9.12}$$

$$= \quad d(x', y'), \tag{9.13}$$

on $\mathbb{R}^2$, and

$$d(x, y, z) \quad = \quad \sqrt{x^2 + y^2 + z^2} = \sqrt{(x')^2 + (y')^2 + (z')^2} \tag{9.14}$$

$$= \quad d(x', y', z'), \tag{9.15}$$

on $\mathbb{R}^3$, between points. In contrast, to illustrate what is meant here, there are geometrical spaces in which the distances between points are not preserved by certain coordinate transformations, such as

$$x' = ax, y' = y, a \gg 1, \tag{9.16}$$

which is a transformation that subjects the original coordinate points (x, y) to shear, but where distances in the new coordinate system (x', y') might not be preserved. Finally we mention briefly that there exist even geometrical spaces such as in Einstein's Special Theory of Relativity, in which the distance between two points is preserved after coordinate transformations[3], even though these particular coordinate transformations can change both the lengths of objects and the duration of timed events!

There are different approaches to the development of good robot motion planning algorithms, after which the algorithms can be implemented by some programming language such as Python, C, C++ or Java. The grid approach superimposes a coordinate grid atop the configuration space. When the number of degrees of freedom is small such as three or six one can implement a search for the best optimal path. With many degrees of freedom there has been success with statistics based sampling methods to find the best paths. The artificial potential fields approach imagines that the point robot is like an electron moving inside an electric potential where there are negatively charged particles while it navigates its way toward a distant positive charge.

So with abstract mathematical approaches such as this one can model successfully how a wheeled robot like a robot vacuum cleaner, should move around in a hotel room without colliding repeatedly into mirrors, walls or closet doors, for example. Additionally with motion planning algorithms there are other robots that either walk and move like humans or manage to negotiate their way successfully around physical objects.

---

[3]In Special Relativity these are called "Lorentz" transformations, with isometry group $SO(3, 1)$

9.3.2 Atlas, BigDog and Asimo

Atlas was the brainstorm of the pioneering robotics engineers who work at Boston Dynamics (www.bostondynamics.com) but also at NASA's JPL, Foster-Miller, Inc., and at Harvard. The robot was developed for entrance into the US Defense Department's DARPA competition. Nearly six feet in height and over three hundred pounds, this humanoid metallic robot, which underwent different physical and technical transformations over recent years, was developed for the military to participate in search and rescue missions and for disaster control operations. Equipped with the latest in Lidar and stereoscopic camera technology for the navigation, physical mapping and reconnoitering of its terrain, Atlas is able to drive vehicles, operate power tools, move debris away from disaster areas, climb up and down stairs, manipulate fire hoses and to turn pipe valves off and on for whenever the need should arise. this machine still is undergoing improvements. Eventually it could be deployed immediately after disasters both natural and manmade presumably such as major fires, hurricanes or earthquakes and for disaster control after terrorist related events.

BigDog (See the Wiki link for BigDog in the Preface) is one remarkable achievement in quadruped based robotics. Its on board computer ran an embedded QNX real time operating system (written in C++) with a truly efficient mechanism for CPU scheduling and memory management. A real time embedded microkernel such as this enforces strict time limits on all processes and threads and manages system interrupts so well that the risk of faulty behavior like deadlock or race conditions is minimized. BigDog had Lidar technology, stereo camera vision and a ring laser gyroscope for inertial guidance via rotation detection. The robot could carry over three hundred pounds of heavy load up a steep incline of thirty degrees, and the quadruped robot could negotiate its motion successfully across various terrains smooth or rough, even across slippery ice.

BigDog used a fifteen horsepower go-cart engine, hydraulically powered actuators and more than fifty sensors to help the robot to sense ground conditions to gauge its movements across different terrains rough, icy or smooth. Unfortunately for the DARPA mission it was the engine that was BigDog's undoing, since the military deemed the machine, which truly was innovative and ahead of its time, still was too noisy for use in the military arena. I admit bias when I claim it is a shame the BigDog project was scrapped, since there must be other significant uses although unknown at present, to suit the astounding abilities of this machine.

As we learned in a previous chapter, Ryerson University's goodwill robot Hitchbot came to a sad and shameful end when it was vandalized by malicious human beings somewhere in Philadelphia back in early 2015. Fortunately Honda's humanoid robot Asimo (http://asimo.honda.com) has enjoyed a more favorable reaction as a goodwill robot.

In Japan Asimo, designed and built by robotics engineers at Honda, has garnered enormous success and it has enthralled many robot lovers and curiosity seekers worldwide in London, New York, Disneyland, South Africa, Canada and Australia. Asimo not only can recognize objects and people, it can distinguish one object or human being from another. It understands voice commands and can tell when a human points to something. It walks, runs, dances, fetches drinks for humans, gives directions and speaks in English and Japanese. At the present time it would appear, or perhaps here I am just biased, that Asimo is about as close one has gotten so far to the compelling and fascinating robot C3PO in emphStar wars.

9.4   Robots that Crawl, Hop, Fly or Swim

9.4.1   RiSE, the Wall Crawling Robot

RiSE is a wall crawling robot built by developers at Boston Dynamics in cooperation with other robotics researchers at Carnegie Mellon, Stanford, Lewis and Clark University and the University of Pennsylvania. Just under one foot in length, this six-legged robot is able to use small claws on its mechanical feet as well as sensors to climb vertical distances along trees, walls and the sides of buildings. One can visualize how such robots could be used in counterterrorism efforts, were it possible to mount small surveillance cameras on the device. Such a robot for example could climb up the exterior wall of a building and directly beside a window under cloak of night, to monitor criminals or terrorists who have imprisoned hostages within an interior room.

9.4.2   Pogo Hopping Robots

Robotics developers at Mit's Leg Lab have developed robots that balance themselves as they hop about on one, two or four legs. This is a noteworthy achievement if one can assume correctly that during any given hop the robot ought to be able to stabilize itself some way from moment to moment to avoid too much pitch, roll or yaw, so that its center of gravity does not deviate too far from some vertical axis of symmetry.

9.4.3   Robots that Fly, Swim, Climb and Crawl

Engineering students at Delft University of Technology in the Netherlands have developed RoboSwift, a bird that mimics real birds in flight with the ability to sweep its wings forward and backward. When cameras are mounted on the underside of the flying device this allows for ground surveillance. This makes such a robot suitable for the monitoring of terrorist or criminal activity such as drug smuggling.

Robots have traversed land and air, so why not the sea as well? Indeed developers familiar with Microsoft's Robotics Developer Studio (this was a Microsoft software package to assist developers who were working on robotics simulation projects) know that this application enabled software developers to build robotic simulations of autonomous boats, even for robotic submersible ships such as the Subjugator which we learned about in the previous chapter. Underwater drones and boats operated by remote control of course are nothing new; in fact remote control boats go back in time to one of the experiments of Nicola Tesla. But other roboticists have gone further than this by developing robotic fish that swim underwater just as do trout, marlin, swordfish and sharks.

One such automaton is RoboFish, a swimming robot that can have applications in surveillance. This automated fish prototype, developed by scientist Oscar Curet and his engineering team at Florida Atlantic University, actually resembles a real fish not only in appearance but also in that it swims underwater as do real fish. Curet did a detailed study on how the Knife fish maneuvers underwater with its fins and in ways that might seem counterintuitive, not only swimming forward through the water using its fins like natural propellers as do most fish, but also exploiting its ability to dart abruptly and suddenly backward and in horizontal and vertical directions perpendicular to its forward motion.

Funded by the US Office of Naval Research, the research on RoboFish that Curet and his team have been doing at FAU has required an amalgamation of different scientific and engineering disciplines–robotics, oceanography, fluid dynamics, computer science, image processing and analysis, to enable the robot not only to resemble physically a fish, but also to learn how to make robots move less mechanically and more naturally at least in an emulated sense or through technological mimicry, and this accomplished through a close examination, observation and study of the various locomotive abilities of fish and even birds.

Another autonomous swimming robot has been developed by scientist David Zarouk and other engineering researchers at Ben Gurion University of the Negev. It goes by the name SAW, which stands for "single actuator wave-like robot." This robot does more than swim underwater, however. It also can climb vertically between certain kinds of walled enclosures such as the interiors of narrow pipes or tunnels by moving like an advancing, highly amplified sine wave. One version of the SAW model was able to crawl horizontally across a surface with a velocity a little less than two feet per second, much faster than the movement of an earthworm, slug or even some species of North American snakes.

This energy efficient, swimming, climbing and crawling SAW robot is minimalistic in its design, meaning that technologically it is not as intricate as is Asimo or BigDog. Yet the autonomous device, with simple parts that are 3D printed actually, is adept enough and versatile enough to be used for surveillance, for reconnaissance or for search and rescue missions. At the time of this writing the SAW robot can be viewed in action at a video posted on YouTube.

9.5 Swarm Robots

We conclude this chapter with a brief mention of other robotics development projects that utilize technological ideas based on mimicry, but in particular animals that engage in various forms of collective swarm behavior. Honeybees, ants, termites and lemmings for example, engage in this type of swarm behavior. As one can tell simply from an online visit to YouTube, there are videos that show different kinds of swarm robots in operation even within a university campus engineering laboratory.

Engineers working on the COCORO[4] (an acronym for *Collective Cognitive Robots*) robotics development project in the EU have been working on a swarmbot that mimics the behavior of schools of fish. Whoever has seen schools of fish in action knows how all the fish in the school can synchronize and choreograph their individual motions almost perfectly, particularly when the entire school is confronted by predators such as dolphins or killer whales. Just as these natural fish are individually able to share and to update information on their environment at a moment's notice, so too the individual swimming robots in the collective COCORO swarmbot share their information with the whole collective.

As robotics engineers continue to study human and animal locomotion they will continue to develop more and more versatile robotics systems, with future applications which only time alone will reveal.

At the time of this writing readers who wish to learn more about these projects can find further information either on YouTube or at www.robohub.org.

---

[4]More information on this aquatic swarm robot project in the EU is at: http://cocoro.uni-graz.at/drupal/home

Chapter 10

Bioengineering and Bionics

10.1 Six Million Dollar Men and Women?

She's breaking up! She's .... We can rebuild him. We have the technology. We can make him better than he was.

Most Baby Boomers who came of age through the nineteen sixties and nineteen seventies will remember the popular TV suspense drama called *The Six Million Dollar Man*. The TV series was based on the science fiction thriller and novel *Cyborg*, by Martin Caidin.. The spin-off series was called *The Bionic Woman*.

Dedicated patriot, astronaut and pilot Colonel Steve Austin succumbs to the violent crash of his experimental HL-10 airplane. After he loses an eye, his right arm and both legs he succumbs to feelings of hopelessness and depression until his mutilated physiology is rebuilt anew in a top secret, six million dollar government research project in bionics. Retrofitted with two bionic legs and arm along with an extraordinary new hi-tech eye, he suddenly can run as fast as sixty miles an hour, leap over tanks and cars with nimbleness and can see out of one eye with an amazing magnification of twenty to one, even in the dark with his eye's infrared imaging ability.

Allied with government supervisor Oscar Goldman in the Office of Scientific Intelligence, Steve and Oscar combat master criminals, terrorists, hostile spies and other different assortments of very nasty villains.

The remarkable thing about *The Six Million Dollar Man* was not just its imaginative, science fiction aspects. More importantly the producers and writers for this TV drama displayed an amazing capacity for science prognostication. Its story line and weekly plots served in a sense like technology's prophetic arrow pointing to the near future. Back in 1973–1978 the expression "We have the technology," spoken by the character Oscar Goldman at the beginning of each episode might have been pointing into the future rather than at the present. But even today the science of bionics has become very advanced in technology, so much so that it makes the much older, body powered (i. e., such as forced muscle contractions provided by the wearer) artificial limbs that were based on mechanical pulleys, sockets and joints, made usually from metal, wood or plastic parts, seem ancient and primitive in comparison.

10.1.1 Machine-Human Fusion

The senseless brutality that was left behind in the wake of the Patriot's Day Marathon bombing on 2013 in Boston, Massachusetts, included brutality that was inflicted upon those runners who were maimed physically as well as those who were murdered. Marathon runners that day lost their limbs due to an irrational religious extremism that motivated and was in turn motivated by the anti-Western bigotry and hatred of the attackers. Yet despite all the grief and tragedy that had emerged from that eventful day, there was still an opportunity for those who lost their limbs to reconstitute their lives anew, in part through the modern engineering science of prosthetics.

Back in 1957–1960 and sometimes in the nineteen sixties my father and mother would travel from Boston to New York and this at times was a frequent event. On such trips I, my parents and sister would stay at the Brooklyn home of a family we had befriended. He was a disabled veteran from World War Two (as my father had been) who lived in New York with his wife and daughter. This veteran had lost an arm and a leg. He had no sophisticated prosthetic devices to aid him in his physical activities so frequently his wife and daughter helped him with his day to day tasks which included among other things, helping him to put on or to take off his clothes, and assistance also with his daily meals. Certainly he could have benefitted from the use of a prosthetic arm and leg. For him at the time however this was not an option. I did appreciate though even as a child how hard it could be whether a Boston Marathon survivor or not, for a person with missing limbs to take on the simplest of tasks, like buttoning a shirt or walking down a flight of stairs.

MIT biophysicist and biomechatronics engineer Hugh Herr also knows, one can assume, a great deal about such a challenge, as no doubt my family friend in New York had known as well as those maimed survivors of the 2013 Boston Marathon bombing. Herr was and still is a rock climber who lost both his lower legs in the White Mountains of New Hampshire when he and a friend were stuck in a blizzard (in the Great Gulf) that caused the temperature to fall almost as low as thirty degrees below zero Celsius. The result of this tragic incident was severe frostbite and the subsequent loss of his legs. Herr did not endure this tragedy alone at the time. His friend also lost toes and fingers and one of the rescuers was killed due to an avalanche.

For a world in which war, human cruelty and uncertain physical events are as commonplace as air, Friedrich Nietzsche's phrase *Was nicht mich umbringt, macht mich stärker*[1] is applicable in many different human contexts. Neither my family friend back in the Brooklyn of the nineteen fifties, nor those maimed in the Marathon bombing or Hugh Herr allowed physical disabilities to define completely their limitations. The family friend did not live the rest of his life as the permanent prisoner of a Brooklyn apartment. Herr continued to do rock climbing, and a dancer who had become an amputee after the Marathon bombing never stopped dancing, with the help of modern bionics.

Bionics more or less is the science of designing and developing technology that mimics things found or observed in nature and also enhancing this attribute when needed, while Biomechatronics is an engineering science that combines biology, physiology, mechanical engineering and electronics. These new engineering disciplines have replaced metal hooks and clamps with bionic hands, artificial hands that have five dexterous artificial fingers that can be moved at will. Wooden legs and feet have been replaced with modern hi-tech counterparts that have enabled amputees to dance, play basketball, run and swim. Biomedical and bionics researchers at the MIT Media Lab, at companies like Mcopro (Medical Center Orthotics and Prosthetics. Visit www.mcopro.com) and elsewhere are doing cutting edge prosthetics research that only is a few years behind the kind of bionics technology that made the fictional character Steve Austin run at sixty miles an hour with

---

[1] "What does not kill Me makes Me stronger." From *Die Göttzendämmerung*, or *Twilight of the Idols*.

artificial limbs.

At the Swiss Federal Institute of Technology and at the Sant'Ana School of Advanced Study in Italy, have made progress in what has come to be called "haptic technology," by inventing an artificial finger that has electrical sensors so sophisticated it can help an amputee to distinguish between a smooth surface and a rough surface. This technology allows the artificial finger to communicate with electrical signals inside nerves within the amputee's arm, instead of utilizing human muscular contractions as was done with the older prosthetic devices of the past. At Johns Hopkins University a prosthetic arm has been developed that helps the patient use actual neuron generated signals to move artificial fingers. In the near future we might see prosthetics advance to such a level that former amputees will have little to no difficulty in distinguishing an artificial leg, knee, foot, arm or hand from a real one.

Not only have there been incredible improvements over the decades in the design and construction of artificial limbs. There also have been scientific and technological advances to assist those who have limited human hearing and sight. Already there exist hearing aids so small when in use that an observer has no inkling whatsoever that the disabled person to whom he or she is talking is wearing one.

Recall that in *The Six Million Dollar Man* TV drama Colonel Steve Austin was given a bionic eye? Today the science fiction of the nineteen seventies has become reality. In at least one case in the United Kingdom an elderly blind man who had suffered from macular degeneration was given a retinal implant to enable him to make out video images attached to his glasses. The technology, developed at Second Sight, Inc. (www.secondsight.com) is called the Argus II Retinal Prosthesis System.

There also is considerable research being done these days in the areas of neuroscience and on what is called "neuromorphic" engineering, to bring sight to the blind. One pioneer in this field is Stanford engineer Kwabene Boahen and the members of his research team, who have been working on a "silicon retina."

10.2 Teleoperated Robots

Users of the Unix operating system, especially those who back in the nineteen nineties, might have read the *Student Guide to Unix* by Harley Hahn, ought to be familiar with both the X Window System and the difference between a terminal and a distant workstation. The X Window system allowed the user to use different applications within separate windows. One could use the Emacs editor in one window, run a Pascal or Fortran 90 program in another and read emails in yet a third window. Yet what the terminal really did was to allow the user at the terminal to use or to create files and applications at the remote workstation. Actually the terminal just was a device for user input and output while the remote workstation did all the real work.

Teleoperated robots at least in principle are similar to the old Unix based computer terminal with its X Window system and workstation systems with which I had become familiar back in the eighties and nineties. The human operator is able get the robot to perform tasks remotely from the location of the operator, just as a Unix user could compile Fortran programs by typing on a terminal keyboard to execute various command line functions, while the actual compilation, program execution and number crunching were handled through a remote workstation. Likewise a teleoperated robot accepts input and output from a distant human user or software application. One example of a robotic system with teleoperation features is NASA's Robonaut, a dexterous robot that has been functioning aboard the International Space Station since 2011.

Amputees such as we have considered in a sense have become cybernetic systems or cyborgs, as was the Steve Austin character originally in the novels of Martin Caidin. Could some sort of future hybrid, a cybernetic coalescence or fusion that is both human and machine, lead to a more stable and productive world, or to a species far superior to both human and machine? Can robots or human-machine cyborgs perhaps, given the right AI programs, be better stewards of the Earth than we humans have been? Humans might not be able to control their violent passions, yet machines have no violent passions to control. In Asimov's novels *The Caves of Steel* and *The Naked Sun* it was the humans, with their homicidal tendencies and various prejudices, who were by far more dangerous than were their robots.

There already are cybernetic systems where humans and machines complement each other to behave as one system. You do that every time you send a fax, drive an automobile, board an airplane or use your smart phone to download a webpage. The future though offers other cybernetic possibilities though, provided the human species does not wipe itself out this century. In the future cybernetic systems and cyborg technology might be so advanced that a human might be able to download a webpage and see the results in his or her mind, or drive a distant automobile remotely with mental commands or with physical gestures, all without leaving the comfort of the backyard patio on a warm Saturday afternoon.

Already there has been work on Brain Computer Interface systems that allow a paralysis patient to manipulate a remote robot arm, and to enable human volunteers to control and to operate miniature quadcopters remotely through the power of the human mind. At Braingate.org engineers, neuroscientists and other scientists have had considerable success enabling paralysis patients, with the aid of a Brain Computer Interface (BCI) system, to get a remote robot arm to reach for a cup. Arrays of electrodes attached to the motor cortex emit neuronal signals that are converted by a computer into control signals, which then move the remote mechanical arm. It is conceivable that this remarkable technology one day when it is developed and advanced more fully, will give a quadriplegic the ability to cook dinner with teleoperated limbs located in the kitchen while he or she sits several feet away inside the living room!

Not only can artificial limbs however, be manipulated remotely by means of BCI technology. It is possible as well to manipulate and to control machines remotely located, perhaps even far away, from the human teleoperator. At the University of Minnesota Professor Bin He (http://helab.umn.edu/) has been able to get human volunteers to operate remote miniature helicopters solely through the power of human thought, at the Biomedical Functional Imaging and Neuroengineering Laboratory. The researchers have adapted the BCI approach to allow brain to computer communication without the use of electrodes connected physically to the human motor cortex. The volunteers instead wear caps on their heads that have electrodes that transmit and amplify signals from the human volunteer's brain so that these can be transmitted as EEG signals to functional magnetic resonance imaging (FMRI) technology. The final stage is for these signals to be converted into feedback and logical control signals output by computer to the quadcopters. Professor He and his colleagues have used mathematical techniques that combine time-frequency analysis, wavelets, principal component analysis and independent component analysis. Time-frequency analysis and wavelets are mathematical "thingies" that are familiar to electrical engineers. Wavelets are used among other things, for data compression. Principal component analysis, used frequently by people who work in the areas of data mining and machine learning, is a mathematical means (from a branch of mathematics known as linear algebra) by which one can reduce the dimensionality of a large data set by filtering out any extraneous or redundant data, when the data set is treated as if it includes points in a random vector space. The Landsat images, so vital in the areas of Geographic Information Systems (GIS), forestry, agriculture and urban planning, frequently combined images from many different remote sensing sources–optical, thermal and infrared–for multispectral image processing. Then this would be subjected to filters and enhanced to remove any redundant noise or data from the final image. Independent component analysis is a mathematical technique that also is important in the areas of signal and image processing. In particular it is used to remove correlations that might exist between data points within a random vector space, to obtain signals that are statistically independent from each other. This is why sound engineers can use special listening devices along with other electronic devices, to filter out ninety nine different conversations to focus on precisely one conversation in a room, such as at a cocktail party.

Will humans one day pilot aircraft from the comfort of a home living room or outside patio? Time will tell where this new cybernetics frontier leads us.

The potential and uses for telerobotics are both widening and profound. Already telerobotics has revolutionized the health care industry. For example InTouch is a company that uses telerobots to treat health care patients remotely (http://www.intouchhealth.com). The Special Purpose Dexterous Manipulator (called Dextre) at the International Space Station is a set of two robot arms capable of making repairs around the station while they are controlled remotely by human operators back on Earth. NASA's Robonauts are humanoid robots, at least for the upper torso. They have been designed to perform repairs remotely at the International Space Station. NASA's HERRO project (Human Exploration Realtime Robotics Operations) points toward a future development of more telerobots in space. Outside the field of space science telerobots have been asserting their presence elsewhere here on Earth. There are telerobots with desktop machines such as those at Revolve Robotics, Inc. (http://www.revolverobotics.com) that can be used to communicate with humans and by showing the remote human operator's image. Other telerobots are built as drivable machines. Telerobots are used by law enforcement and for videoconferencing as remote avatars. Some future applications could be for remote administrative and secretarial services, for remote cooking in the food services industry, to operate sewing machines remotely or even to perform remote medical checkups or surgery.

Finally we mention here briefly about *haptics*, that area of telerobotics that involves transmission of information that conveys to the remote user something analogous to human touch. Decades ago game users got a sense of this when they used a steering wheel in a virtual reality environment. If the virtual road got rough or bumpy the user got a feel for it when the steering wheel was beset with vibrations. With haptics one can get a sense of the "feel" of a fabric when the robot holds the fabric somewhere miles away from the human teleoperator. A medical student remotely can "sense" when the scalpel he or she holds cuts into some virtual kidney.

Some of these technologies however require more than electrodes, actuators and other hardware. Frequently software also is required, this particularly is true for the autonomous robots we have considered previously, such as BigDog and Atlas. In order to advance beyond the very simplistic feedback and control or input-output systems of the past, robots need to learn in order to adapt and to update information.

To do this autonomous robots indeed do need software and special algorithms for the software to help them to run artificial intelligence and machine learning programs. These two topics we treat in the next chapter.

Chapter 11

Machine Learning, AI And Robot Immortality

11.1   Knowledge Discovery in Databases

A *database* is an organized system for storing, maintaining and retrieving data records. Whereas in ancient Babylon data records might have been kept on cuneiform tablets, we kept data records usually on paper years ago, as documents that were stored in filing cabinets before the advent of the computer age. Suppose back in December, 1953 a sales manager at a New York or Boston department store wished to know what a particular customer named Henry Glockenwinkel might have purchased from April, 1950 to July, 1951. If we assume they used no data processing systems at the time, than more than likely he or she would tell a secretary or office clerk to go to a file cabinet and look for "Glockenwinkel, Henry" within some listing of inventory or sales files all appearing perhaps under the initial "G." Depending on how small or large is this paper file system, the human powered search might take two minutes or twenty minutes of time. Such long times were cut short with the arrival of computerized database systems of the fifties.

One such computerized database system was the hierarchical database and one of the most famous examples of this kind of database was IBM's Information Management System (IMS) which was run even on mainframes way back in the nineteen sixties. Another example of a hierarchical database could be found at the old fashioned Gopher protocol text document websites back in the early nineteen nineties.

This kind of database however is not obsolete. Today Microsoft's Windows Registry is an example of a hierarchical database. Even XML documents are arranged as a hierarchical database, in which information is nested within other information, all connected by links so that the whole thing resembles an inverted tree of sorts. Today both C and Lisp programmers are familiar with something called "linked lists," a data structure in which the "nodes and branches" of the tree can be searched in a downward and upward manner, as in Lisp. Computer scientists call this type of search a *linear* search, because it can be done in linear time[1], in contrast to polynomial time.

Suppose we want to find the number nineteen located somewhere inside a listing of

---

[1]Mathematicians and computer scientists usually denote linear time as $O(n)$, read as "Big-oh of n," which means the computational search can be completed in computer time that is bounded above by some positive integer n multiplied by some positive constant.

numbers. We have given to us a large list of numbers,

$$1, 8, 5, 65, 89, 4567, 2, 3, 4, 25, 37, 196, 19, 33, 45, 76, 89, 0, 1, \cdots \qquad (11.1)$$

For the search the computer sets the variable x to $x = 19$ then initiates the search, using some algorithm and setting some variable $y[i]$ equal to each one of these given integers one at a time for each $i = 0, 1, 2, \cdots$, checking each time to see if the condition $x == y[13]$ is met when $i = 13$.

Today there are other kinds of databases in use, in particular what one calls relational database management systems. These systems have a special data definition language (DDL) and data manipulation language (DML) called SQL, based on the relational algebra and relational calculus research in the nineteen fifties and sixties by IBM computer scientist Edgar Codd (1923-2003) and on what was learned about the formal languages we discussed in a previous chapter. With database systems such as these, one needs to perform multiple queries with minimal cost in contrast to computing some algorithm in linear time. Multiple queries, that is, queries that are all included in batch processing in a relational database system, might not be done with minimal cost if the system is not designed to optimize the queries, for example if the database tables contain too much duplicate or redundant information in them. The same customer might have moved repeatedly over the course of several years of orders and they might have changed their phone numbers several times. If the database stores all different addresses and phone numbers for each and every customer that moved and if there are tens of millions or hundreds of millions or more of these, it could increase the cost considerably. In fact some RDBMS's are enormous in size. They can contain hundreds of thousands or hundreds of millions or even billions of data records! For such large database systems some multiple queries might not get done with minimal cost, especially many queries that join together data from millions or hundreds of millions of different database tables and where some of these database tables might have redundant information. In fact finding ways to optimize queries in a relational database management system has been an active area of relational database research for many years.

Today we have relational databases for different purposes. Every time you purchase a pair of sunglasses or a book online from a large ecommerce website, your purchase leads to the sharing of your transaction data at some backend database server which one would hope of course, is secure. When there are millions of customers making transactions every day the backend database frequently must update records and the data does change frequently. Such a database system is called an *online transaction processing system*, or OLTP system.

Another kind of relational database system is an *online analytical processing system*. This kind of system stores data for online use and for the purposes of data mining. A "relational algebra" is a formally precise, description that also is an abstract mathematical model that demonstrates the operational procedure that exists behind an SQL query in a real working, relational database management system. Suppose a database administrator works at some IT company which develops IaaS cloud based applications for clients. This DBA needs to "Retrieve the employee names of all the company software developers who have worked for the corporation for the last five years, who are earning more than ninety thousand dollars a year and who work in the IT department for the content management system developers." Then the names, the years these employees were hired and the salaries are all attributes in an EMPLOYEES table and the names and IT departments in which they work are attributes in the DEPARTMENTS table. Then the DBA might use an SQL query like

```
SELECT NAME
FROM  (SELECT NAME, YEAR_HIRED, SALARY, DEPT_NAME
       FROM EMPLOYEES NATURAL JOIN DEPARTMENTS
       WHERE SALARY > 90000 AND YEAR_HIRED ='2011'
       AND DEPT_NAME ='CONTENT MANAGEMENT';);
```

One can express this in the relational algebra, using for example symbols σ for SELECT and $\bowtie$ for JOIN. The relational algebra details what steps the SQL code should take to output something, just like an algorithm tells a computer program how to compute something to get its output. However one cannot use the relational algebra to output real data any more than one can use an algorithm written in pseudocode to output data.

There is a second abstract formalism called the "relational calculus", which has two versions, one of which is called the tuple relational calculus. Both the relational algebra and the relational calculus are formal languages like those formal languages we addressed in a previous chapter, and the relational calculus is very similar to the predicate calculus used in first order logic.

Now the point is we can borrow the relational algebra to form graphs that actually are trees, that model or depict query searches in a database. A tree that corresponds to our query appears in the Figure.

If the query search is very complex in that is has for example nested queries with lots of join operations, then we also can model such queries with a query tree. Moreover we can use the abstract query search tree for relational databases to find those optimized queries that will be the most cost effective for any organization's operations, when they have to use an OLTP system that might be dealing with heavy load or scalability issues, when it comes to the daily use of multiple database queries. There is a machine learning algorithm that can be used for this purpose called the genetic algorithm, which we will discuss in an upcoming section.

11.2 What is Data Mining?

We all know what miners do. Coal, bauxite, nickel, copper, magnesium, gold, platinum, uranium, these are precious ores, metals and minerals that miners extract from a mine. Today both science and business need data which can be compared to precious ores. Just as a coal miner prospects for coal, data scientists mine for data, sales data, crime data, census or biographical data. With data mining one extracts valuable data from a RDBMS. Then one cleans and preprocesses the data, that is, to manage errors, missing values, and to make sure the data is consistent with each other since, if some of the data values lie between zero and one while other values are greater than ten thousand, at least some algorithms deployed to understand the data might be rendered useless. In the business world there is a plethora of buzzwords, including in relation to data mining: data munging, data dredging, predictive analytics, etc. Suffice it to say that raw data ought to be preprocessed first before anyone seeks to extract useful patterns of knowledge and information from it, be he or she a data analyst, data scientist or applied statistician.

Scientists and even business managers need to know certain facts. There might be astronomical image data on billions of elliptical galaxies, spherical galaxies, irregularly shaped galaxies and even some stars. An astronomer might want to classify galaxies into two categories like elliptical and spiral. Then too underwriters might need to classify credit card holders into those who are credit risks and those who are not credit risks. Before these things can be done the data first has to be mined.

We mentioned briefly about OLAP systems. One first extracts and transforms data from OLTP systems then loads it onto OLAP systems to complete data mining tasks. Strangely enough the data either can be stored in accordance with the relational database ideas of Edgar Codd or else stored as data "hypercubes"of sorts, that can have millions or even billions of dimensions.

One then processes probability and statistics based "machine learning" algorithms on the datasets in some programming language, say Python or R, to extract patterns from the data that help management or scientists make decisions about the datasets. Such a database system one calls a *Decision Support System* (DSS).

However data mining is not just about the extraction, transforming and loading of data into OLAP systems. In a sense data mining is a part of machine learning.

One ought to mention that there are other data mining platforms that are not based on the relational model, such Hadoop/MapReduce, Apache Spark, so called "NoSQL" systems. Hadoop and MapReduce differ from RDBMS's in that MapReduce does not support multiple reads and writes as does an OLTP system and the Hadoop File System (HDFS) differs considerably so from the way data is structured within an RDBMS.

11.3   What is Machine Learning?

Machine learning involves the use of algorithms and computer programs that implement those algorithms, in such a way that the computer "learns" something it could not have learned otherwise solely from the computer code it runs and compiles. For example a computer program can specify how exactly a computer can find the square root of π and print that out in floating point notation. On the other hand a computer program in general cannot get a computer to predict whether or not one particular online customer will like or dislike a pair of shoes he or she sees displayed in an advertisement on some ecommerce webpage. For one thing the programmer does not know even who the customer will be in advance, or what his or her preference in shoe styles will be.

Machine learning happens when a computer system uses computations based on statistical algorithms, to find knowledge patterns in data. This is in stark contrast to the (START, ACCEPT/REJECT) finite state automatons and finite state machines we discussed earlier. Software systems modeled on ACCEPT, REJECT or YES, NO really are state based systems that depend upon logical "if" Boolean conditions (See Figure 11.1). "If the first name is George and not Thomas, look for last names that begin with T." "If $x = 19$ then PRINT 'the sought for number is in position i,'" etc. Programs that use logical flag conditions like these can be modeled by Boolean circuits and logic gates[2]. This though is not what one would call machine learning. The program just is getting the computer to reach final states, based on what is input. No, machine learning involves the use of statistical algorithms to reach forecasts or predictions from the data, and this either can help the computer system to make decisions such as in a control system, or else in the case with databases, it helps scientists, engineers or business management to make

---

[2]This is true for "classical" computer systems based on logic based on the two results in the set $\{0, 1\}$. Quantum computers though use quantum logic and quantum logic gates, two topics that are beyond the scope of this book.

decisions. This is in contrast to other forms of artificial intelligence such as AI based on symbolic logic, which we shall consider soon in another section.

There are many different kinds of statistical machine learning algorithms to get this

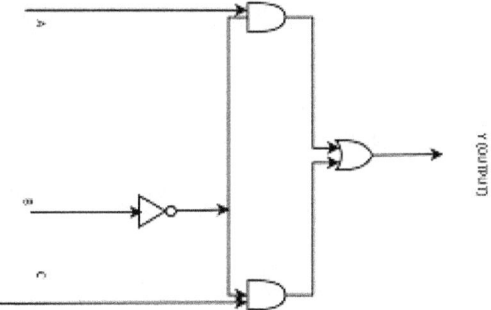

Figure 11.1: An example of a Boolean circuit.

done. Usually one puts them into two very specific groupings, supervised learning and unsupervised learning algorithms.

Take supervised learning for example. Data analysts know beforehand that they can put credit card applicants into two categories, those with a good credit history and those with poor credit ratings. So one knows beforehand the data will fall into these two groupings and patterns then are found about age, education level, income level, demographics, etc. This type of data usually has "training sets" with which one can use supervised learning algorithms. Classification and decision trees, regression models, support vector machines, these only are the names of some statistics based supervised learning algorithms that enable data scientists to detect patterns in data.

Unsupervised learning on the other hand means one knows little about the patterns because they lie hidden in the data. K-means clustering, Gaussian mixture models and artificial neural networks are the names of some algorithms for unsupervised learning. One more example is anomaly detection which is very useful for detecting suspicious behavior in a computer network, like remote unauthorized login attempts. Then one can use also algorithms such as Markov models and Hidden Markov models, two algorithms which among other things, are used today in computer network security to check the strength of passwords and in computational biology, to filter out the background audio noise that gets into spectrograms for the ultrasonic vocalizations of mice.

Still there are other machine learning algorithms that help one to optimize something or to come up with a very good approximation, like in finding a very good approximation to minimizing the value of some global cost function, like the cost function we considered in regard to relational database queries. One example of this is called a *genetic algorithm*. The algorithm derives this seemingly peculiar name because it describes behavior that has been modeled by researchers in mathematical biology, who have modeled how complex chromosomes are built up over time with genes and alleles that undergo mutation during the process of evolution and natural selection.

Consider again the sought for objective or cost function for relational database query optimization. We need to minimize this, which is what the genetic algorithm does. Just as chromosomes are built up by genes that mutate over time according to those working in mathematical biology, so too by the genetic algorithm a global cost function

$$\sum_{j=1}^{k} f(v_j), \tag{11.2}$$

is modified over time by "mutations" to derive the query that best approximates the query that minimizes the cost function. Here v_j, where $j = 1, 2, 3, \ldots, k$, denotes some binary string of bits $\{0, 1\}$ that gets "mutated" whenever some bit in the string gets changed and where each bit string like this one is a single node in the search tree[3]. For instance the string 1000100 in the algorithm might mutate to the string 1000110, which yields locally for each j a better approximation $f(1000110)$ than does the previous one at $j - 1$, if $f(1000110) < f(1000100)$. The various mutated bit strings in the algorithm correspond to the genes that undergo mutation in the chromosomes. When each bit string v_j that best optimizes the overall cost function is found, a sum of discrete probabilities

$$\sum_{j=1}^{k} p_j, \;\; 0 \le p_j \le 1, \tag{11.3}$$

gets very close to the value one, which means the approximation that is the best fit to the optimized query has been found.

11.3.1   IBM's Watson

One fascinating example on the success of machine learning has been Watson. Watson, a question answering system developed at IBM by Tony Pearson and David Ferrucci, is an incredible and awe inspiring assemblage of software and hardware built to be a participant on the TV game show *Jeopardy*. Its data storage capacity is formidable. It used different database platforms to store both structured data and unstructured data. It can access data from its storage in the terabyte range with its Hadoop File System and it performs teraflops and teraflops of computations. Unlike the kind of heuristic search tree algorithm that helped to implement the Samuel Checkers Player, Waton uses an information retrieval system to find out how to make correct answers that are input to it verbally by a human.

The high level Watson software architecture has several subsystems or subcomponents, one subsystem or subcomponent for initial question management and for query decomposition and text parsing (natural language processing), another for information storage, a third for information retrieval, a fourth for hypotheses formulation (i. e. perhaps either with automated theorem proving or an inference engine of sorts) and synthesis and one for answer generation. It has at its command data from millions of books, newspaper articles, encyclopedias and other sources of written text. But unlike online search engines, its information retrieval system did not output page ranked documents after having received some keywords. Instead it parsed the questions input to it, searched and sifted through reams of data until it found what was pertinent. Then it would output the answer in spoken words, depending upon the keywords it had found in the initial question.

In 2011 Watson answered most of the *Jeopardy* game show questions superbly well,

---

[3]In actual practice the real "tree" treats SQL queries, particularly nested queries, as if they are trees derived from the relational algebra and that might have several join operations.

just as well as did the human contestants, delivering its answers in the calm, rational speaking voice given to it through text to speech software. Watson was a different sort of artificial intelligence from what researchers had developed generations earlier, like the Arthur Samuel Checkers Player and Deep Blue, which we shall consider shortly.

11.4 John McCarthy and the Development of Symbolic AI

Remember some of the artificial life forms, automatons and robots we considered previously? Whereas both *La Joueuse de Tympanon* and the Mars Pathfinder simply were automatons devoid of all intelligence, Tik-Tok, the Frankenstein monster, Gort, the HAL 9000, the replicants in the film *Blade Runner* and the Federation starship android named Data all possessed an extraordinary ability. They could think, reason, use logic to reach decisions and speak with humans, and they were conscious of their own existences. In the realms of science fiction and horror they possessed intelligence although it was artificial, since they either were machines or things created through medical science. But if any one individual was the originator of true artificial intelligence so that it exists today within the realms of computer science both pure (i. e. the actual AI theories) and applied (i. e. the actual algorithms, code and robots to implement AI or machine learning), that person would have to be John McCarthy (1927–2011).

Even after his graduate school days in mathematics at the California Institute of Technology and at Princeton, John McCarthy must have possessed a seemingly boundless amount of intellectual energy. As either both professor or researcher he was associated with no less than four universities: Princeton, Stanford, MIT and Dartmouth. Not only was he the forerunner of AI research; he also was a promoter of the use and exploitation of computer timesharing systems[4]. McCarthy also helped to design and to develop the ALGOL programming language. He invented an early version of "garbage collection," a technique by which a computer programmer can manage the way they allocate computer memory and free memory[5]. This is important because one can develop a software application that includes millions and millions of lines of software system code, so that if for example there are too many memory leaks caused when the code runs as a process it could slow the process down considerably.

In 1956 John McCarthy also was instrumental in organizing the Dartmouth Conference. This conference and workshop was a groundbreaking, seminal event in the history of theoretical AI research and development. In fact it was John McCarthy first who coined the well known expression "artificial intelligence." Some of the other attendees at the first Dartmouth Summer Research Project on Artificial Intelligence included Allen Newell, Claude Shannon (whose research on information theory and entropy we learned about in a previous chapter), Oliver Selfridge, Herbert A. Simon who was doctoral advisor to Allen Newell at Carnegie Mellon and Marvin Minsky.

One of the attendees at the 1956 Dartmouth Conference was Allen Newell (1927–1992), a researcher in computer science and cognitive psychology with a background

---

[4]This Author had actual exposure to such timesharing systems, back in the days of computer tape libraries and then later running FORTRAN batch programs on a mainframe, with computer punch cards and keypunch machines, in the nineteen seventies.

[5]Both $C++$ programmers and Java programmers can do this in two different ways. In $C++$ however it is handled directly in the code more or less by the programmer.

in physics and mathematics. Newell did research work on organizational theory at the Rand Corporation along with Herbert Simon, focusing on logistics problems for the US Air Force. For a time he collaborated with mathematician Joseph Kruskal, who was an early and active participant in the American Civil Rights movement. Some student science students might be familiar with his algorithm, which is related to the minimum spanning tree in a weighted graph[6].

Herbert A. Simon was not a computer scientist by profession. Actually his interests lay more within the world of large organizations such as corporations. Particulary he did research on how organizations make decisions and found that large organizations have many components and subcomponents to them. A large airline has flight reservation systems, a personnel department, security, baggage personnel. All these different departments do not share the same interests. Those in personnel want to hire and schedule the best pilots. Managers in charge of baggage wish to cut down on any baggage that has been misplaced or lost. Security at the airport desires to conduct surveillance to cut down on the possibility of theft or other criminal or suspicious behavior. Still others in the airport's organization are in charge of drafting the flight schedules. But what happens when the effeciency in one department depends upon the efficiency in another department? Such dependencies between different departments has an impact on what sorts of decisions must be made for the overall efficiency of the airline. Simon believed that to solve such problems one must go beyond mere mathematical or algorithmic approaches. Back in the nineteen fifties this led him to develop an interest in computer simulation and in the burgeoning new discipline of AI.

Allen Newell developed two important AI programs, one of which was for information processing, and eventually like John McCarthy he won a Turing Award for his contributions in artificial intelligence research.

One of the other attendees at the Dartmouth Conference was Oliver Selfridge (1926–2008), who completed his undergraduate studies in mathematics at MIT. He pursued graduate studies there with Norbert Wiener as his advisor, although he did not complete the Phd program. Selfridge[7] made important contributions in the areas of machine perception and pattern recognition.

Many AI researchers at the time, in particular those we just have mentioned who were in attendance back in 1956 at the Dartmouth Conference, were familiar with Lisp. John McCarthy had invented this programming language. He modeled it on what became known as the λ calculus, which was invented by Alonzo Church.

11.4.1   The λ Calculus, Lisp and Symbolic Programming

Many of these aforementioned early researchers and pioneers in AI, that is, John McCarthy, Oliver Selfridge, Allen Newell, Herbert A. Simon and others, put great emphasis on symbolic logic over statistics based machine learning. Perhaps that does seem reasonable to some extent. Neither a two year old human toddler nor even a real tortoise named Elsie uses computation by statistics to make decisions. Both adapt through experiential learning and by having the ability to store new information. A toddler by experience engages time and again neuronal processes to associate a table or chair with the word "table" or "chair," that represents the concrete object in actual human language.

---

[6]One example of a "weighted graph" would be the graphs used in operations research for network flow models.

[7]Enthusiasts of PBS *Masterpiece* ought to be familiar with the show *Mr. Selfridge*.

Early in his career in fact the philosopher Ludwig Wittgenstein had considered as did Ernst Mach, the possibility that logic and empiricism are the basis for much in physical nature even more than is metaphysical reasoning. Even Kurt Gödel during his studies at the University of Vienna had warmed to this idea.

Mathematical logic is based heavily on two foundations, propositional logic and the predicate calculus. Students in discrete mathematics know only too well how to work out truth tables for propositional implications like

$$p \rightarrow q, \tag{11.4}$$

The predicate calculus much like the relational calculus we considered previously, almost seems like a language by itself, with expressions like

$$x, m, b \in \mathbb{R}, \exists y \in B \subset \mathbb{R}, \ni y = mx + b, \tag{11.5}$$

which in English can be read as, "For every x, m, b in the set of real numbers $\mathbb{R}$, there exists some real number y in a subset B of $\mathbb{R}$, such that $y = mx + b$."

Perhaps in part at least, for this reason the early researchers in AI believed symbolic logic was ideal for deriving models as to how computers one day will be able to think and to reason. But there is something else here to consider. recall the work we learned about earlier, that people like Alan Turing, Alonzo Church, Stephen Kleene and others were doing with regard to computability and Turing machines? At the time they had placed emphasis on something called a primitive recursive function.

A recursive function is analogous to those nested Russian dolls that get smaller and smaller when you open one of them. To illustrate with the factorial function,

$$
\begin{aligned}
f(0) &= 1, & \text{(11.6)}\\
f(1) &= 1, \\
f(2) &= 2 \cdot f(1) = 2 \cdot 1 = 2, & \text{(11.7)}\\
f(3) &= 3 \cdot f(2) = 3 \cdot 2 \cdot f(1) \\
&= 3 \cdot 2 \cdot 1 = 6, & \text{(11.8)}\\
f(4) &= 4 \cdot f(3) = 4 \cdot 3 \cdot f(2) \\
&= 4 \cdot 3 \cdot 2 \cdot f(1) & \text{(11.9)}\\
&= 4 \cdot 3 \cdot 2 \cdot 1 = 24, \\
\cdots &\quad \cdots & \text{(11.10)}\\
f(n) &= n \cdot f(n-1) = n \cdot (n-1) \cdot f(n-2) = \cdots = n! & \text{(11.11)}
\end{aligned}
$$

There are different kinds of recursive functions considered in theoretical computer science, such as general recursive functions (K. Gödel did some research on these), primitive recursive functions, etc. However more details about this is beyond the scope of this book.

Modern computers also can use algorithms that deal with recursive expressions, like

```
If (BOOLEAN == 1)
     Y = 1;
     X = ((Y*Y) + Y + 1);
Else
     Y == 0;
END
```

An algorithm can behave much as does a finite state automaton or as does a finite state machine if it produces output when something is input into it. The opposite also is true, meaning a finite state machine can behave much as does an algorithm. In fact the Church-Turing Thesis asserts that a Turing machine behaves very much like a certain kind of function that is computable by an algorithm. By a "certain kind of function" we mean one that is computable, recursive and defined on a finite subset of the natural numbers, meaning it is a "partial function." This has to be the case because some recursive functions can be computed and run by computer in a finite number of steps. For instance if a finite state machine tries to reach a final state for

$$f(n) = n^n, \forall n = 1, 2, 3, \cdots, 100^{100}, \cdots, \tag{11.12}$$

it never will halt.

A Turing machine of course uses a tape with symbols on it and eventually it reaches a HALT state if the computation is to be like an "effective procedure" that ends after a finite number of steps. So to characterize its behavior Alonzo Church invented the λ Calculus. The λ Calculus is an abstract kind of symbolic programming language that models the behavior of a recursive function that can be used to compute something in a finite number of steps. Now John McCarthy evidently was so impressed with the λ Calculus that he invented an actual symbolic programming language based on it called Lisp[8].

Lisp was the *lingua franca* of the AI community from the nineteen fifties up until the nineteen eighties. Even today it enjoys a resurgence among some computer programmers in versions like Common Lisp and the functional programming language Scheme. Today the powerful text editor named Emacs still is popular among the Unix crowd of enthusiasts[9]. It is a text editor written in both Lisp and C.

Lisp has some very powerful features that make it extremely useful for the development of artificial intelligence related software like expert systems, although some programmers loathe the language, complaining about its seemingly bewildering plethora of parentheses. For one thing Lisp, like the functional programming language Erlang, allows a programmer to create functions that behave like a black box. This is not true necessarily in other typical structured programming languages or for object oriented programming languages. Although for instance C, Perl and Java have been used for functional programming, usually this involves functions and methods that either involve the initialization of declared variables or else instantiation. With Lisp one creates functions that input things then outputs things, using S-expressions and even recursion whenever necessary. This actually allows the programmer to get his or her program to modify code within the code, without resorting to statistical methods, to perform various artificial intelligence tasks.

With Lisp one can create automated theorem proving systems, software systems that can use symbolic logic and either modus ponens along with forward chaining and backward chaining, or else "resolution refutation," to reason out the answer to complex questions input to it, such as, "Joe is a man and a husband and Amanda is a woman and a wife. Joe and Amanda are married and Harry is their son. Meg is Joe's aunt and Fred is Amanda's father. Is Fred Harry's grandfather? Is Meg Harry's grandmother?"

Moreover for AI symbolic programming with Lisp can be more beneficial than conventional programming languages in which variables get declared then initialized. In

---

[8] It stands for *LISt Processing*.

[9] The Author had opportunity to learn Emacs on a Unix system, as an undergraduate student in mathematics.

comparison some high level programming languages like Fortran and C are ideal for solving serious "number crunching" problems in particular on parallel or high performance computing systems, for instance for finding solutions to linear programming problems in which there are millions of unknown variables and constraint equations, and for which the feasibility region for the solutions lies inside and on some convex polytope with millions of dimensions. But although "mother," "father," "children," "son," "daughter," "aunts," "nephew," "grandson," "grandparents" and "grandfathers," can be defined as data types such as in C^{10}, Lisp has much more powerful symbolic programming features for answering the complex questions in the previous paragraph.

So if a problem has to do with fruits and vegetables, say, Lisp as a symbolic programming language can be used to answer fruit and vegetable related questions, by associating for instance "apples," "oranges" and "bananas" with "fruit" and "cabbages," "radishes," with "vegetable." With the right Lisp code a program can answer questions like "Is an apple a vegetable," "are oranges red" or "Are radishes yellow?" Now suppose a robot in a supermarket was programmed with Lisp code. It is conceivable it could be trained to reach for an orange instead of for a radish when it is told by a human to do so, then to put the orange into a paper bag for a customer, a task that need not be completed with computational or numerical based algorithms.

In most programming languages, one might want to evaluate a definite integral with a program, such that the output is given, using pseudocode, as

$$Integral(0.0, 0.5 \cdot (\pi), x, sin(x)) = 1.0, \tag{11.13}$$

with output 1, when with symbolic programming one might want instead the expression

$$\int_0^{\pi/2} sinx dx = -cosx + C, \tag{11.14}$$

as output from some question answering system that is being used to teach calculus to a high school student.

Another property that Lisp has is that it uses linked lists as data structures (See Figure 11.2). This accounts for the use of so many parentheses in a Lisp program. With Lisp the programmer can move up and down a tree structure, for instance when he or she needs to do a binary search.

By using Lisp, its powerful S-expressions feature and functional programming, AI

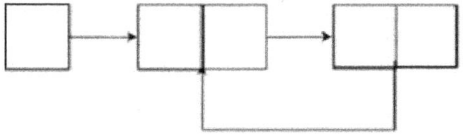

Figure 11.2: A Circular Linked List.

researchers were able to design, code and test programs to create rules based systems or inference engines, to solve capacity planning and logistics problems, to build expert systems and to engage in machine translation. The Dynamic Analysis and Replanning

---

[10]In C this might be done with the use of structures, which in C is something like a public class without methods (such as in object oriented programming), and by using **typedef**.

Tool (DART), a software application used by the US military in operation Desert Storm, was based on Lisp. So was SPIKE, a scheduling software system used for the Hubble Space telescope.

11.4.2   Strong AI

In one of his seminal papers Alan Turing addressed the issue as to whether or not computers will be able to think in the same sorts of ways that humans think. His argument boiled down to a paradigm that has come to be called The Turing Test. A human sits outside a room and uses a teletype to communicate with someone or something in an adjoining room. The human asks questions and the person in the adjoining room uses the teletype to respond to each and every question. If the responses make sense to the interrogator and seem to be the kinds of sensible and sequential responses one would expect from a human being, the interrogator assumes there must be another human in the other room. Strong AI asserts that one day, machine intelligence and even things like sentience, self-awareness and consciousness will be possible with machines. Others who accept the Strong AI idea opine that any algorithm, program or computational procedure has a kind of mental process at work behind it, whether it be the procedure by which a thermostat functions, a computer or the human brain. Some other people however outside the field of symbolic AI though took exception to this position.

One such person who took exception to the arguments of "Strong AI" researchers, was the philosopher John Searle. He had a powerful counterargument against the argument that machines one day will be able to think as humans do and to be conscious as humans are conscious.

The counterargument actually is a *Gedankenexperiment*, or "thought experiment," like the kinds of thought experiments Albert Einstein devised to represent his arguments. One calls Searle's counterargument The Chinese Room. British mathematician Sir Roger Penrose discusses Searle's counterargument in his book, *The Emperor's New Mind*.

One version of the counterargument goes like this. There is a native speaker of Chinese who sits outside a room. This native Chinese speaker has cards with Chinese characters written on each of them. The native Chinese speaker puts the cards in some ordered sequence of arrangement so that they spell out or convey some intelligible sentence, story or message in Chinese. He or she then slips them in the ordered sequence of characters through a slot inside the door that leads to another room.

He or she cannot see who or what is in the other room, but in the other room there is a man who speaks English but does not know Chinese. The room also has boxes that contain various Chinese written characters, each character written on a separate card, say. The room also contains a special book of instructions on how to select Chinese characters from the boxes in an ordered, sequential way and to output them through the slot in that ordered sequential way, to respond to the other party who had input the Chinese characters into the room in ordered sequence. The ordered sequence of Chinese characters the human operator within the room outputs to the human outside the room after following the instructions in the book, forms the sensible written response to the other party's question, although the human inside the other room does not know Chinese. Since the ordered sequence of Chinese characters that are output from The Chinese Room makes sense to the native speaker on the other side of the door, this native speaker assumes there is a human being inside the room who understands Chinese.

Now imagine that the book inside the room that gives instructions to the human operator inside the room as to how exactly to choose for a reply the ordered sequence of Chinese characters from the boxes, is an algorithm or computer program. Furthermore the boxes that stack or hold the cards with Chinese characters on them, either is a database or some kind of memory storage. Searle then concludes that the book of instructions that denotes the algorithm or program for the correct response in Chinese characters does not know or "understand" Chinese any more than does the human operator sequestered within the room. This is The Chinese Room argument against the position of Strong AI.

There are at least two powerful rebuttals to this reasoning. The one on which we focus here states that one should not expect either the book of instructions or the human operator inside the room to understand Chinese. Yet if one looks at the entire *system* that comprises The Chinese Room, then Searle's counterargument fails to hold for such a system. Indeed the system itself *does seem to understand Chinese*. The book has the right instructions on how to respond; the human operator then "knows" how to choose the proper characters from the boxes that serve as the correct response to the input.

For instance a school or "swarm" of fish does seem to "know" or "understand" how to avoid a predator such as a dolphin, although one individual fish in the school or swarm might not understand the overall significance of its actions. But is this a fallacious stance? Perhaps not, unless someone correctly proves otherwise.

Nevertheless research in AI went on the wane particularly in the nineteen eighties, sometime after the development of Lisp. Some claim that the corporate business world, with its endless profit margins, bottom lines, tax havens and "buzzwords," hijacked AI away from the research community of computer scientists. Others have claimed that symbolic languages like Lisp were much too slow to use for AI applications in comparison to other high level programming languages like C, such as for the development of expert systems. Some people in AI began to feel that computational or statistical approaches were better to resolve AI issues rather than lists and S-expressions. Nevertheless over time symbolic programming and symbolic logic proved not to vanish away from AI research any more than Fortran had vanished from work in scientific programming, after the development of C, C++ and Python.

11.4.3   The Samuel Checkers Player

Arthur Lee Samuel (1901–1990), an electrical engineer who completed his studies at MIT, truly was a remarkable study in adaptability and versatility. Like the company IBM at which he worked for many years his research and development goals were subject to frequent adjustment and change, to keep pace with the rapid and extraordinary transitions within the world of technology as it bridged the gap from the heat accumulating vacuum tubes and low memory analog computers in the nineteen forties to transistors and RAM.

In his book *Mind Matters: Exploring the World of Artificial Intelligence*, James P.Hogan describes how and why Samuel left Bell Laboratories back in 1949 to work at IBM. After World War Two the technology scene was changing, transforming into an ever expanding, brave new world. So back in 1949 Samuel chose wisely not to be stuck permanently inside the anachronistic world of vacuum tubes and electromechanical computers.

Since his interests in programming and AI were diverse and variegated so were his contributions. As a programmer he was an innovator in contrast to being merely a developer of software for some business application. In the field of artificial intelligence one of his most important contributions was his checkers player.

Checkers is a simple enough game to play but it can be deceptive for new players who are unwary. Back in the nineteen fifties I was one of these unwary players who lost one checkers match after another frequently to my own war veteran father, who often was bemused by my youthful bewilderment each time I lost. Yet much later I learned the more complex game of chess from a schoolmate back in 1970 and had opportunity some years later to teach my father the game.

Games like checkers and chess–called "combinatorial games" by computer scientists and game theorists–is a zero sum two person, fair game: *Winner take all, Loser gets nothing.* The number of possible outcomes for any one game each time two opponents sit down to play checkers or chess is an enormous number although that huge number is bounded above and finite. Some people might be misled to think there is an infinite number of ways one checkers or chess game can turn out, but such is not the case. In fact computer scientists know that the computational complexity for checkers in terms of computer time is EXPtime-complete. Basically this means that for a checkers game a deterministic Turing machine can accept or reject (decide) in $O\left(2^{8^2}\right)$ time, or in time bounded above by 2^{8^2}.

It was not until the nineteen nineties that a truly formidable computer chess player named Deep Blue arrived on the scene to play Gary Kasparov. Until that moment arrived however, the Samuel Checkers Player gave excellent insight as to how a computer with the right program for game playing can "learn" from its mistakes. The Samuel program, which he started to develop when he was at the University of Illinois, ran on the IBM 701. Basically its algorithm generated search trees in which there were nodes that represented the moves of each of the two players. The initial tree, in which each node had subsequent nodes that represented all the next moves that were possible for it, was updated frequently so that subsequent trees would be subject to tree branch "pruning," to remove all the other moves that a piece either could not make anymore or did not make.

The most advanced version of Samuel's checkers playing program made its debut in 1955. It learned to play checkers and to win against different opponents, even against itself on occasion, through feedback, trial and error in a heuristic sense, updating to new information when the need arose. It was a kind of rote learning as it decided how to advance to the next best position. A linear "scoring polynomial" kept track of the program's best move at every step across the board and the program would backtrack the steps it had made to get back to its initial position at the beginning of the game, summing the optimum value of its linear approximation function. If this value was the largest possible the program had made to the best move along a particular branch of the search tree. An additional feature was the use of a numerical value called a *ply*, by which the program could keep track of all the best moves it was making by keeping the ply value at a minimum.

At any current position in a match the heuristic search tree denotes all the nodes for the program's moves to the current on-move position and that of the opponent's to his or her current on-move position[11]. A "backup" algorithm then retraces the on-moves along each possible branch that leads from the possible current on-move positions for each of the two players, back to the initial root (See Figure 11.3). This way the program decides

---

[11]For a more detailed description one should consult Chapter 10 in the book *Mind Matters: Exploring the World of Artificial Intelligence* by James P. Hogan we mentioned previously.

which on-moves are best, by comparing the on-moves along the different branches. If the program's on-move positions have been poor all the way up to its current on-move position, then it compares the other on-move possibilites along the different branches that lead to different current on-move positions, to the backup to find what the branch would have been to an on-move with the best possible overall moves. But if its on-move positions have been good all the way up to its current on-move position, then that particular branch was optimal.

A *piece advantage* used by the program was a benchmark of sorts. It measured how

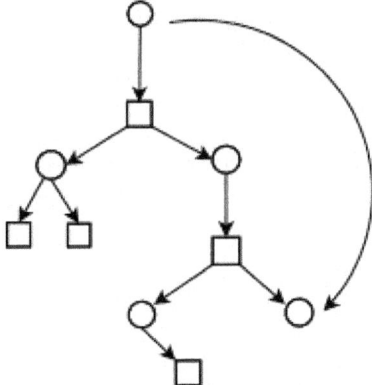

Figure 11.3: Search tree and "backup" (See arc with arrow). Circles are the program's moves, squares for the human player. All the moves not shown.

good the program was playing by the number of pieces and kings it had remaining on the board.

The Samuel Checkers playing program was able to beat many an accomplished amateur in checkers matches. It became a milestone in the development of computers running AI programs that play games, including Deep Blue and later after that, DeepMind.

11.5   Marvin Minsky

As we have seen already in Part I there were two giants of science fiction, namely Arthur Clarke, with his artificially intelligent HAL 9000 series spaceship computer and Isaac Asimov with his humaniform robots on the planet Dawn. These two science fiction authors in particular seem to have done more than anyone else to popularize the possibilities of what robots and intelligent computers can do. But just as they popularized groundbreaking conceptual ideas on robots and artificial intelligence, Marvin Minsky as well as John McCarthy were the two correspondoing giants in the real world of artificial intelligence research.

Minsky though did not seem as interested as was John McCarthy in any future progress for symbolic AI. His approach took him in a more computational direction, in particular with his ideas on "perceptrons."

It was John McCarthy who, for the first time, used and defined the term "artificial intelligence." It was Marvin Minsky though, whose name became associated far and wide to this field of computer science right along with that of John McCarthy. Like many scientists who chose to do research in AI from the nineteen fifties onward, Minsky seems also to have been a 'Jack of all trades, master of all' of these professions: Mathematician, electrical engineer and computer scientist. This certainly does seem appropriate though, since theoretical computer science as a pure *science* did have two parents, pure mathematics and electrical engineering. From mathematics it derived the theory of formal languages, database theory, complexity theory and the theory of computation and compiler development, with which Grace Hopper was so familiar. From electrical engineering computer science derived flip-flop gates, switching circuits, memory registers and computer hardware.

Since both as a scientist and researcher in artificial intelligence Marvin Minsky had a professional background not only in mathematics and computers but also in electrical engineering, one can add the trade "inventor" to his listing of professions which the well liked Professor Minsky (MIT) had mastered. As an inventor Minsky designed a very sophisticated confocal microscope, as well as the Logo turtle with Seymour Papert, and SNARC in 1951 with Dean Edmonds.

What was SNARC?

SNARC is an acronym for Stochastic Neural Analog Reinforcement Computer. In essence it was one of the earliest electronic prototypes for an artificial neural network that uses "reinforcemnt learning." Unlike an artificial neural network simulation built up within the confines of some programming language, SNARC was a real machine built with the state of the art technology for the 1950s: A gyropilot (taken from a World War Two fighter bomber) and hundreds of clutches, valves and motors, with electrical wiring connections which Minsky and Edmonds chose deliberately to randomize throughout the device's electronic circuitry. Indicator lights also were built into the thing so that as it ran one could observe the overall pattern of neuronal computations as a series of light flashes moving in succession, like laboratory rats moving about, navigating their way through a complex maze.

SNARC had only forty artificial neurons, control knobs turned by the electric clutches. The reinforcement learning was achieved as closed loops, memories that would appear at random through the overall circuitry, thus increasing the probability values around these memory loops in the circuitry to get more favorable outcomes.

SNARC, which had been constructed by Minsky and Edmonds based upon the ideas of Warren McCulloch and Walter Pitts, made quite a sensation among the AI community at Harvard, MIT and even Dartmouth in these early years.

A McCulloch–Pitts network was an abstract agglomeration of logic gates and switches, but not a real physical device like SNARC. Unlike the probabilities that were used in the physical SNARC machine, these abstract McCulloch–Pitts networks had connections with weighted values for zero or one. There were AND gates within it, as well as OR gates and gates for NOT, NOR and subnets for delays, feedback and control. Minsky gives an extensive description of McCulloch–Pitts networks in Chapter 3 in his book, *Computation: Finite and Infinite Machines.*

Minsky realized though that these sorts of artificial neural networks, McCulloch–Pitts networks, really were just another example of a finite state machine with memory. Indeed most computer science undergraduates know that a pushdown automaton does happen to have a memory stack. So in Section 3.5 in his book he presents us with the following converse theorem which we reword here for emphasis, since we are talking here specifically about McCulloch–Pitts networks:

> *Every finite state machine is equivalent to and can be simulated by some McCulloch–Pitts network.*

Recall that we met finite state automata and finite state machines back in Chapter 3.

11.5.1 Perceptrons

Artificial neural networks (abbreviated ANN for short) were conceived in an attempt to model the way electrical information signals cross neuronal pathways within the brains of some biological organisms such as in the brains of humans where electrical signals flow across the synapses between neuron cells. Even though many ANNs have been successful in what they are designed to do, still some AI researchers have questioned that they model accurately how *real* neurons and synapses within the brain process information and store memory. For instance one can note that although ANNs frequently use weighted numerical values for the connections between artificial neurons, this just is not the case with real information that is spread between brain neurons. Put simply real neurons in the human brain for example do not deploy mathematical probability whenever a human shopper is trying to remember where she parked her car in the supermarket parking lot.

This however does not prevent ANNs from achieving the expectations of their human

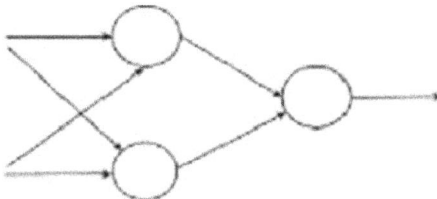

Figure 11.4: Perceptron with one hidden layer (Weights not shown).

developers, such as in the areas of pattern recognition and memory storage. ANNs such as the perceptrons in the two Figures are called "feed forward" because the flow proceeds forward from the input to the output.

In 1957 Frank Rosenblatt worked on the Mark I Perceptron at Cornell University. This early perceptron machine was being used for image recognition. But back in the fifties and sixties however some did not hold out much hope that ANNs really could be used beyond certain narrow areas of application.

In their book *Perceptrons*, Marvin Minsky and Seymour Papert outlined in detail some of the limitations of perceptrons. One hidden layer was not enough for such a perceptron to be very useful. Also they argued that such perceptrons could not handle XOR operations. What is an XOR operation? There is one for logical propositions, one for sets and one version for combining two bit strings. That is, with logical statements P and Q, a Perceptron was not much use for handling logical expressions such as

$$P \oplus Q = (P \vee Q) \wedge \neg (P \wedge Q). \tag{11.15}$$

With two nonempty sets A and B,

$$A \oplus B = A \cup B - A \cap B. \tag{11.16}$$

Typically also it is an XOR encryption algorithm one uses for the secure block encipherment of "one time keypads," made by combining two different (binary) bit strings with an XOR computation. Recall that we already considered the possibility that one can have securely encrypted one time keypads in Chapter 6, where that chapter mentioned Claude E. Shannon's paper "Communication Theory of Secrecy Systems."

ANNs, in particular those ANNs that one calls Hopfield networks, are capable of reinforcement learning. Under certain conditions whenever two nodes that are connected together change their values within a Hopfield network, the weighted values along their connections (edges) can increase and in this way the two nodes are associated. When this occurs the weighted values along the connection are "reinforced." If this happens in the right way between nodes within the network so that local minima are reached for a certain energy function, the energy value is decreased, which means a minimizing of the error between some predicted value and some estimated value for some supervised learning problem. Now to do all this the connected nodes or units as they also are called simulate a logic gate or switch. Let n_i and n_j be two such connected nodes where $i \neq j$. For some computations one requires that either n_i or n_j is in the ON position, say, but not both of them. The other must be OFF. That is, the Hopfield network or perceptron must be able to resolve computations like

$$1 \quad \oplus \quad 1 = 0, \tag{11.17}$$
$$1 \quad \oplus \quad 0 = 1,$$
$$0 \quad \oplus \quad 1 = 1, \tag{11.18}$$
$$0 \quad \oplus \quad 0 = 0. \tag{11.19}$$

But if the perceptron in question is a single layer perceptron (Figure 11.4 is an example of a single layer perceptron, while Figure 11.5 is another one that might not have enough dimensions for XOR.), there could be some activation functions or threshold functions, such as perhaps the Heaviside unit step function

$$H(t) = \begin{array}{ll} 1, & \text{if } t > 0 \\ 0, & \text{if } t \leq 0 \end{array}$$

in some situations, so that it would not be able to resolve, model or simulate computations which requires the use of XOR operations. This would be highly relevant for instance, if the related real world problem was to use the perceptron to resolve some pattern recognition problems.

Suppose such a single layer perceptron with insufficient dimensionality (See Figure 11.5) or that did not have the right activation function had to distinguish between handwritten letters L, F and T but could not resolve any logic based XOR operations. Let

$$A = \{L, T, F\},$$
$$B = \{L, F\}.$$

Let j be the index for the jth neuron, k, $k = 1, 2, \ldots, m$ the index for the kth input value x_k arriving from across a synapse and w_{jk} the weight along the connection link to neuron n_j. For this j^{th} neuron the single layer perceptron needs to compute a scalar product of two vectors

$$net_j = \sum_{k=0}^{m} x_k w_{jk},$$

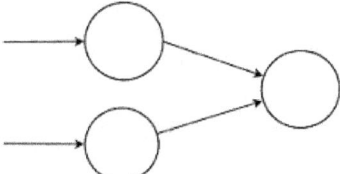

Figure 11.5: Perceptron that could violate linear separability (Weights not shown).

where the two vectors in question are

$$X = (x_0, x_1, x_2, \ldots, x_m),$$
$$W = (w_{j0}, w_{j1}, \ldots, w_{jm}).$$

When computations for all the neurons, input values and weights are included the perceptron needs to find a correct output value y for some overall threshold function

$$y = f(net).$$

But if the single layer perceptron cannot evaluate XOR expressions properly, such as for example by following some logical rule like the one in Eqn. 11.15 (along with the use of certain threshold functions), it might have a problem trying to recognize the handwritten letter "T" correctly in the set

$$A \oplus B = \{T\}.$$

It would be as if the perceptron was unable to follow an instruction rule, perhaps something like "Find the element in the set A OR in the set B, AND NOT in the set A AND in the set B." In order to do this the solutions would have to be divided properly by some decision boundary or boundaries, into two or more different classes determined by something called "linear separability."

Some have claimed the conclusions that Minsky and Papert had reached about single layer perceptrons helped to put the brakes on artificial intelligence research particularly in the nineteen eighties.

So Minsky and Papert's contention held that perceptrons with only one layer were

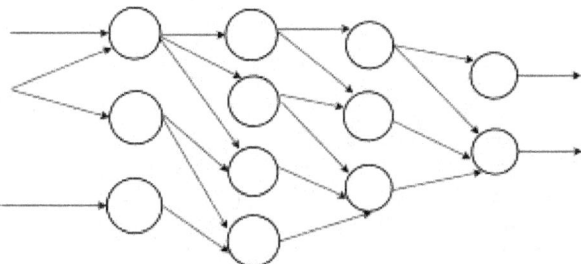

Figure 11.6: Multilayer Perceptron with three hidden layers (Weights not shown).

not sufficient for most perceptrons to do more sophisticated forms of computation.

But the seeming abandonment of AI did not last indefinitely. After computer processing power increased AI research returned as well along with research on new kinds of ANNs, such as Hopfield networks (See Figure), Boltzmann machines, convolutional neural networks and recurrent neural networks, ANNs which today use many more than one hidden layer. Today they have many different layers of neuronal pathways and connections (See Figure 11.6). With the vast increase in the number of hidden layers these newer ANNs were capable of widening their uses.

Today within the corporate business world there are a plethora of IT buzz words:

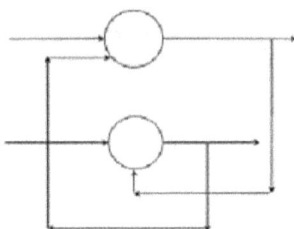

Figure 11.7: Hopfield Network with one hidden layer (Weights not shown).

Smart phone, business intelligence, knowledge management, knowledge engineer, recommender systems, predictive analytics, data munging, to name a few. One can add another one to the list after the great success of artificial intelligence in general and ANNs in particular, things which already have been exploited widely to create new consumer products: Deep Learning.

One recalls a line from one of the couplets of Alexander Pope: *Drink deeply from the Pierian spring.* Pope meant that if one wishes to learn do not restrict oneself to a shallow level of learning. When one studies a foreign language for example such as Italian, Hindi or Cantonese and if one wishes to read works of literature in these languages, it does not suffice to restrict one's learning to some small foreign language pocket guide for international travelers. One needs to learn the grammar, verb forms, sentence structure and idioms, the nuances, etc. Something similar is true for ANNs. If an ANN or perceptron is to be of any widespread use it must be capable of deep levels of learning. This is why some of the most sophisticated ANNs today have, not just one or two hidden layers with a few nodes, but hundreds of thousands and even billions of hidden layers and nodes.

Today this is the case with deep learning research at Google. Deep learning and reinforcement learning are far more complex artificial neural networks than are the simpler multilayer perceptrons we considered. For one thing reinforcement learning ANNs can have millions of neurons and connecting links. Moreover instead of being feed forward mechanisms they utilize bias, feedback and back propagation so that the numerical weighted values along the connecting links frequently get updated.

Google scientists in deep learning have been doing highly significant research on ANNs that have mind boggling numbers of artificial neuronal connections within parallel systems for image and speech recognition. In the past ANNs barely could distinguish a silhouette of a dog from that of a cat. Today deep learning ANNs have moved far beyond such archaic abilities, capable not only of distinguishing dogs from cats but also spiral galaxies from spherical ones. There also have been important applications for deep learning artificial neural networks in medicine, pharmacology and molecular biology. With the use of reinforcement learning ANNs can be used to transcribe a spoken speech into text with minimal error and in small amounts of time.

The potential for deep learning is vast and the application domains for its use are extensive. With the right sets of training data taken from surveillance cameras, customer invoices, online cookies and travel records, law enforcement working in homeland security can use deep learning systems capable of deploying pattern recognition, sentiment analysis, machine translation, fuzzy logic or Boolean satisfiability, to forecast the possibility of a terrorist attack.

On the American television drama *Person of Interest*, brilliant IT inventor and physically disabled billionaire Harold Finch and his shadowy CIA veteran friend John Reese used their clandestine access to a gargantuan government AI computer system called The Machine to save the lives of potential victims who were targets either for future criminal acts or for future abuse from corrupt government officials. The AI computer system did this in part by using an extraordinarily vast geographic information system (GIS) replete with a spatial database and machine learning and AI algorithms that Finch had designed. One only can admire these two human protagonists for being superheros without spandex or capes, as they used AI along with the clandestine help of a police detective, to save the lives of the innocent instead of using AI to take those lives or to ruin them.

Combined with virtual reality and augmented reality systems or haptics technology, deep learning systems could enable medical students to diagnose disease or to simulate surgical operations conducted on a virtual human patient, then predict the outcome for healing. Foreign diplomats and business executives could travel to a foreign country and use a portable machine translator or foreign language translation and transcription device to communicate with people overseas, without the need for a human translator.

11.6 More Complex Games

Deep learning ANNs can be used to enable a computer to play games that are much more difficult than are checkers and chess in terms of finding winning strategies, such as the game Go. To date however no one has programmed a computer yet to solve the Tower of Hanoi puzzle in tractable time.

The Tower of Hanoi would be quite the challenge for either an AI program on a computer or a robot to solve[12]. As the mathematician Édouard Lucas describes it, this game involves the use of a wooden board on which there are three adjacent vertical pegs. One of these pegs has n rings stacked on it for $n \geq 1$, rings that decrease in diameter as one moves upward along the stack of rings on the peg. The object of the game is to stack all these n rings onto exactly one of the other two pegs so that the moved rings also decrease in size from the bottom to the top on the other peg to which they are moved.

One only can move one ring at a time to the other peg. One cannot place a larger ring onto a smaller ring on a peg. Also one is free to use a third peg as a placeholder to hold one ring as one moves another larger ring from one peg to the next. If there only is one ring then obviously one can move this solitary ring to one of the other two pegs in a minute amount of time. If there are two rings one can do it in no less than three moves. With three rings one can do it in no less than seven moves.

For each integer $n \geq 1$ there is a graph associated with the possible moves. The graph

---

[12]I first learned of this game when I was an undergraduate in mathematics at the University of Massachusetts.

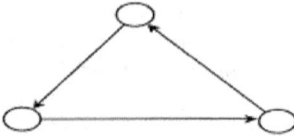

Figure 11.8: Graph when the number of rings is $n = 1$. The actual graph is undirected. The arrows indicate the possible peg selections.

for $n = 1$ is just a triangle with three nodes and three edges, where the nodes are the three possible positions for some number of rings on each peg and the edges are the moves taken between the three pegs (See Figure 11.8). One can take the solitary ring from peg one and put it on peg two or take the ring from peg one and place it on peg three or leave it where it is. For $n = 2$ the graph is a triangle with three smaller triangle subsets (Figure 11.9). For $n = 3$ one has a much larger graph structure made from nine smaller triangles and that reminds one of the Sierpinski triangle fractal structure. Figure 11.8 depicts the graph for $n = 1$ ring and Figure 11.9 depicts the graph for $n = 2$ rings. Usually these graphs are undirected, meaning they normally are not graphs with directed edges. Here the edges are arrows only to indicate to the reader the moves that are possible as one plays the game with n rings. For instance when $n = 2$ take the smaller ring from peg one and place it on peg two, then the larger ring from peg one and place it on peg three, then take the smaller ring on peg two and put it atop the larger ring on peg three. Or take the smaller ring from peg one and place it on peg three, then take the larger ring on peg one and put it on peg two, etc. Thus for one of the triangles in Figure 11.9 the distribution of rings would be the larger ring on peg one while the smaller ring either is on peg two or on peg three or else both still are on the first peg. This makes up the three possible nodes and the three possible edges for the triangle on top in Figure 11.9. As n increases each peg at any given time could have more than two rings on it.

For any positive integer n one can make the correct transition of rings from one peg to one of the other two pegs in less than $2^n - 1$ moves. So for any fixed large positive integer n_0 one can ask a question, namely,

QUESTION: How large an ANN is needed (i. e. in terms of the number of nodes and connections needed) to complete the Tower of Hanoi puzzle in no less than $2^{n_0} - 1$ moves?

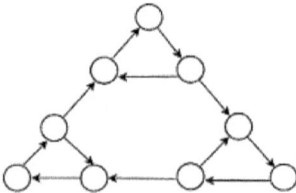

Figure 11.9: Graph when the number of rings is $n = 2$. The actual graph is undirected. The arrows indicate the possible peg selections.

A very pertinent question one ought to ask is whether or not it is feasible even to try to get some ANN to solve this problem when the number of rings is large. It is known that when n is any large integer, the average number of moves is[13]

$$\frac{466}{885}2^n - \frac{1}{3} + o(1), \tag{11.20}$$

where $o(1)$ just means any tiny real number ε between zero and one. We have seen there have been computers that could play checkers and chess. There also are robots today that can play ping pong with a human opponent. Might the Tower of Hanoi be a challenging game for an intelligent robot that uses a deep learning artificial neural network as it moves the ring from peg to peg? One can wonder if there might be a connection between the number of rings in the game and the number of ANN nodes and connections needed to solve the game.

DeepMind Technologies Limited in the UK is an AI company that achieved considerable success in the development of reinforcement learning ANNs that are capable of learning how to play a wide variety of games of strategy and video games with the use of image analysis. After acquisition by Google in 2014 the scope of ANN development widened to tackle games of strategy as challenging as Go.

Now at Google an ANN dubbed AlphaGo has done this. The game Go is more than two thousand years old. Two opposing players move white and black stones across the game board to capture opposing pieces and to gain territory in the form of unoccupied squares. The number of different strategies are huge in terms of computational complexity. Due to the multiplicity of possible game strategies AlphaGo uses Monte Carlo simulation to tell it which possible moves will maximize its chances at success, deep artificial neural networks and reinforcement learning instead of heuristic search trees. AlphaGo managed to defeat an experienced human player as well as several other Go playing AI programs.

11.6.1 Knowledge Representation

Although we had to digress for a time to discuss the applications of artificial neural networks, we have not forgotten at all about Marvin Minsky. After all it is to Marvin Minsky we owe a debt of gratitude for the recent developments in ANNs, since he truly spurred interest in them with his early research on SNARC.

Yet when it comes to research in AI we owe Minsky a debt of gratitude not only for the expanding area of deep learning, something which has artificial neural networks at its core as did SNARC although this was at a far more primitive level back in the nineteen fifties. We also can thank him for helping to develop ideas on how computers and robots not only can access knowledge, such as knowledge in databases or search trees, but also on how to *represent* it.

Knowledge representation encompasses a means to help an AI system classify a system of knowledge represented as objects and their attributes within some specific domain of knowledge, such as medicine, online shopping or astronomy. However more is needed than just to have a classification system or taxonomy. An AI machine ought to "know" things like birds, airplanes, bats, ostriches and wings. Yet an effort at classification alone would be useless since the AI system could not tell that ostriches have wings but they do not fly and that bats do fly and they have wings but they are not birds.

Thus the right classification system sometimes is called an "ontology" if it includes as well relationships between different classes, objects and attributes. Decades ago in his

---

[13]See the Bibliography.

paper "A Framework for Representing Knowledge" Minsky introduced the concept of a "frame." Today many different kinds of frame languages and ontologies for knowledge representation, and most of these are for web based applications, such as OWL and the Semantic Web.

Some researchers today in the field of computer science and artificial intelligence might not have forgiven Minsky for his influential paper on perceptrons along with Seymour Papert branding him perhaps, as being some avowed enemy of perceptron research. But research in science always has elicited disagreement and debate.

Regardless of whether or not one disagrees with Minsky's conclusions on single layer perceptrons, the many contributions in artificial intelligence by John McCarthy and Marvin Minsky have been with us since the Dartmouth Conference they attended back in the nineteen fifties. Still we see today thriving communities of active Lisp programmers. Furthermore the steady increase in processor power, thanks to Moore's Law, has not witnessed yet the demise of perceptrons in various AI applications. When one contemplates the range of their expertise and the expansiveness of their ideas along with their contributions to AI which certainly have not been insignificant such as Minsky's ideas on frames and perceptrons and McCarthy's ideas on symbolic logic and the many uses of Lisp, few I think would argue against the assertion that it is possible that both Marvin Minsky and John McCarthy have done more than others to shape the future directions in AI research for at least another decade if not for the next fifty years.

11.7   Robot Immortality

Today there is an idea among at least some theoretical physicists and cosmologists (I do not presume to assume that physicists would call it a school of thought), that the universe is mathematics. By that I do not mean to say these scientists believe merely that the universe is mathematical, or described or governed by mathematical laws but that it *is mathematics*. By this idea I mean the physical universe actually is some kind of mathematical structure from which all matter is derived whether photon or matter particle, black hole, planet, spiral galaxy or sentient being. By "mathematical structure" one can assume this to mean a "structure" just as the ring of integers under addition and multiplication is a mathematical structure, or like an isometry group[14] such as those used in quantum mechanics (like the unitary group $U(1)$) or in particle physics (like the symmetry group $SU(2)$ that particle physicsts love to work with) is or is part of a mathematical structure. A higher plane curve studied in analytic geometry, like a limaçon, epicycloid or the cissoid of Diocles is or is part of some mathematical structure all its own. I phrased it this way "is or is part of ..." because for example a *monoid* with its elements and binary operation is a larger mathematical structure than is a group with its elements and its binary operation, so that a group actually can exist within the monoid as a proper subset, a fact that would make it part of the monoid when its elements are considered to be elements of the monoid.

Today the Holographic Principle in physics and the new discipline of quantum computation bolster the idea that the physical universe is computable just as an algorithm

---

[14]Isometry groups figure prominently in the kinds of linear transformations that are important for actuators to engage in certain robot motions like translation or rotation of an axis and for robot maneuverability. Mathematical groups along with quaternions also play a "background role" of sorts, in the software development of inertial guidance and virtual reality applications. The more mathematically daring can find books on group theory in the Bibliography.

in computer science used for a solvable problem is computable by a finite state machine. Through the work of cosmologist Max Tegmark at MIT it has come to be known as the *mathematical universe hypothesis*, and sometimes as the *computable universe hypothesis*.

The computable universe hypothesis is a fairly profound idea. If it is true it might answer some very ancient questions without human recourse to any religious dogma or to superstition to find answers that philosophers and theologians have been asking for millennia. But what if our physical universe truly is an aggregation of various material particles, fermions interacting mathematically and computationally by means of what particle physicists and cosmologists describe as massless gauge fields that interact or even perhaps compute somehow, on the boundary of some discrete spacetime? Quantum computer and quantum computation experts like Seth Lloyd and David Deutsch at MIT describe physics at the quantum level and so all of physics by derivation classical and otherwise, as being determined in a sense by the processing of quantum bits. If these ideas are right it suggests that all the interaction of massless boson fields and the processing of quantum bits determines the very nature of our universe, its physical laws and all the material bulk within the cosmos: Quarks, protons, electrons, neutrons and neutrinos, dark matter, dark energy, Higgs particles, gravity waves, radio galaxies, Cepheid variables, even college professors, the expensive computer science textbooks needed for your next semester, distributed database server networks, laptops and robots. The phrase "All is vanity," has been attributed to the wise Solomon, but if the universe is computable concept is right, it means even the Wisdom of Solomon is computable.

If the computable universe hypothesis is true then all of our material universe was computed by massless quantum bits similar to how classical bits stored on a server determines the actions of the avatars in a computer game.

Some like philosopher Nick Bostrum have suggested it is possible that the universe and everything in it is part of some computer simulation.

To compute atoms and photons, galaxies, planets and stars? That is a lot of computation. Does it mean our physical universe is an intelligent quantum computer? That is a question best left for the skeptics, the philosophers and the religionists to ponder.

What about robots? How intelligent can they become? Can they live forever? Roboticist Hans Moravec gives us some indication in his book, *Robot: Mere Machine to Transcendent Mind*. The book title even demands serious contemplation. Neither the Mariner 9 space probe nor the KUKA robotic arms were transcendent minds. On the other hand Moravec does give us an example of a race of robots that evolves over time from highly sophisticated cyborgs to a cosmic web of intelligence that transcends a mere machine confined to a physical form.

Imagine a future in which some humans themselves, so enhanced spectacularly by biogenetics and through technological enhancements have evolved into superhuman intelligences. Lust, hate, greed, cruelty and war no longer are relevant. They left earth space long ago, since the confines of Earth they found to be too restrictive for their expansive natures to grow. They need the universe itself for their laboratory and playground. No longer do they have the need to continue themselves through procreation. The "Exes" as they are called can replicate themselves and they enjoy the convenience of having become self–repairing systems[15]. Physical improvements are handled through a process of self–organization.

---

[15]Already there has been research on robots that can repair themselves when they have access to 3D printers to print out parts for repair, and swarm robotic systems that repair themselves by exploiting a modular design (See Bibliography).

Countless eons after such superhumans have attained this level they have advanced into a complex system so enormous and interconnected that they have abandoned entirely the personal confines of any physical form whatsoever, to reach a level of advancement so superior that they fill the entire future universe at the level of the quantum vacuum. The universe in the far distant future actually becomes a sort of intelligent web or internet comprised of countless superior minds that exist at the quantum level as they span a complex network topology. Such a species would need technology as we understand it no longer.

By then the former cyborgs would have reached the level of pure collective mind. If such a superintelligent cybermind were capable of being described as being on the Kardeshev scale, they might have reached the level of a Type IV civilization.

It would be as if the entire universe countless billions of years in the future was one enormous, sweeping intelligent neural network, processing all information at the quantum level. All such a superintelligent mind would have to do is to proclaim *Fiat lux*! Then there would be light. At the very least such a scenario can be treated in a science fiction plot as the improbable made possible. Even death itself might be conquered, not by religion, but through a kind of computational resurrection. There are in fact cellular automata systems in which the artificial life forms can "die" but then return to life in a different digital form.

Chapter 12

Robots that Chat and Talk

C'est pas par là! C'est par ici!
From *La Cantatrice Chauve*, by Eugène Ionesco.

"What's a 'tweet'?"

Just imagine the Starfleet android Data asking such a question to a visitor from the twenty-first century!

The American version of the English language is diffuse with grammatical paradoxes, oxymorons and double entendres. Business colleagues say to one another "I'll email you," when the word "email" is a noun. In the workplace and on university campuses one mentions in casual conversation that they "saw something in a tweet." In the nineteen eighties and nineties college friends who encounter each other years later after graduation would say warmly, "Let's do lunch," or "I'm good,"instead of, "Let's have lunch together" or "I'm feeling well." Now just imagine an intelligent machine trying to understand such spoken language usage.

In the past there have been those who have explored the many contradictions and inconsistencies that can arise in human conversation. In his novel *Catch 22* author Joseph Heller confronts us with a plot centered on US Army Air Corps units in which official bureaucratic language and human conversation become a sheer nightmare for US pilots fighting the German Luftwaffe during World War Two. Due to the bewilderment and confusion engendered by a military bureaucracy the novel's protagonist, Captain Yossarian, is more terrified of his commanding officers than he is of the Nazis. Any flyer has the right to lodge a complaint against an abusive military policeman although there is no way to prove the abuse actually occurred. Another pilot is good natured and a decent human being, but others do not like him. A pilot can get a discharge if he is "crazy" but if he recognizes he is crazy then he really must be sane after all.

After World War Two the French playwright Eugène Ionesco did something similar in his absurdist play *La Cantatrice Chauve*[1]. Two married couples in England, a maid and a fire chief have a dinner conversation diffuse with one disconnected discussion after another in which one given line of conversation from one party at the table has no relation whatsoever to the other responses that follow. At the very end of the play there seems to be even a confusion of identities as the play begins afresh with the lines of dialogue that had been spoken by one couple at the beginning of the play are spoken at the end of the play by an entirely different couple.

What is the play suggesting, one might wonder. Does it suggest that when conver-

---

[1] *The Bald Soprano.*

sation between humans is superficial or trivial, one party really is not listening to the words spoken by the others, or that communication itself between humans has become meaningless in a modernist society?

Spoken language communication and meaning has been a topic not only for playwrights and novelists to explore. Recall the MIT SHRDLU program by Terry Winograd we encountered back in Chapter 4. Still another researcher in computer science who was interested in the topic of human and computer conversation was Joseph Weizenbaum.

Only Weizenbaum was much more interested in using a computer program to show just how superficial a conversation between two parties can become.

12.1   ELIZA

Just how well can a computer fool a human with which it is communicating? Can it fool a human into thinking it is a second human being that is making the conversation? This is the question that computer engineer Joseph Weizenbaum wished to consider when he built his ELIZA[2] program.

He wrote ELIZA in MAD-SLIP[3] for the IBM 7094 mainframe, a time–sharing computer system where input was entered to the computer from a remote terminal then the output would be returned by the mainframe. MAD–SLIP was an early programming language that combined algorithm development with data structures called circular doubly linked lists, something very similar to the circular linked list we considered already in Chapter 11 (See Figure).

Weizenbaum's program utilized both a pattern matching and transformation rule mechanism. Each of several keywords were associated with a transformation rule and all these comprised something like a stack or list. It gave ELIZA a remarkable property. ELIZA did not have to "understand" its responses at all. Instead it simply gave responses that frequently made perfect sense to the human who communicated with it through a teletype.

Here is an example of how it worked although here we do not claim this conversation actually occurred. A human could type

`My boyfriend really has me angry today`

and ELIZA would respond

`WHY DOES YOUR BOYFRIEND HAVE YOU ANGRY TODAY`

Then the human might type

`Because he took my car without asking me first`

After which ELIZA could respond

`WHY DID HE TAKE YOUR CAR WITHOUT ASKING YOU FIRST`

and so on. The trick is that the program would respond to a pattern like (where we are using Unix shell regular expressions here to indicate the pattern matching)

---

[2]Named after G. B. Shaw's Eliza Doolittle character we considered previously.

[3]MAD stands for Michigan Algorithm Decoder. SLIP stands for Symmetric List Processor. This was a language similar to LISP.

```
My.*has me.*today.
```

But there are *plenty* of English sentences that can match a similar pattern, such as

```
My dad has me so surprised today.
My stock portfolio has me so worried today.
My test grade has me feeling good today.
My screen test really has me feeling super today.
My dental checkup has me concerned today.
My school soccer team has me feeling so proud today.
My lab experiment's result certainly has me astounded today
```

and so on. So one can use even the Unix shell to find some specific pattern in a string or line of English text then use a shell script to return a second string that seems to respond correctly to the first string, although the shell will not "understand" at all what the line of English text means.

12.2 Modern Virtual Humans

12.2.1 A.L.I.C.E.

The ELIZA program has not been the only program designed to mimic successfully an actual conversation between a human and an intelligent machine. AIML, or the Artificial Intelligence Markup Language, is a subset of XML developed by computer scientist Richard Wallace. With the use of tags, pattern matching and animation software one can create an embodied conversational agent, chatbot or "virtual human" like ALICE[4]. The Alicebot can perform very well within a specific domain of use, such as for online help desk applications. AIML interpreters have been written in a variety of programming languages, such as C and Java. With some ingenuity and practice a skilled web developer can incorporate a virtual assistant feature on a website or in some desktop application. However with the wrong scripting a virtual human's responses to questions can be far removed from any correct responses. At the time of this writing one can visit YouTube to hear "conversations" between two different chatbots.

12.2.2 Ada and Grace

One finds another example of virtual humans with Ada and Grace, two virtual human twins who act as guides for visitors at Boston's Museum of Science.

As a child in the Boston Public Schools system back in 1959–1963 two things left a lasting impression on me when my elementary school class one year visited the Museum of Science. The first was Spooky the Owl, a very imposing, great horned owl that was the museum's mascot, and the museum's Van de Graaff generator.

Back in those Eisenhower and Kennedy years however there was no technology display at the museum that could compare to Grace and Ada[5]. Today these are two virtual humans located at Boston's Museum of Science and maintained under the collaboration of the University of Southern California's Institute for Creative Technologies and Boston's Museum of Science. Grace and Ada are museum guides that also comprise a question

---

[4]Artificial Linguistics Intelligent Computer Entity.
[5]Named after Ada Byron and Grace Hopper (See Chapter 3 and Chapter 4).

answering system for the museum's visitors young and old alike. With a combination of automatic voice recognition and natural language processing algorithms the two "virtual twin sisters" are remarkably realistic in the way they give sensible responses to many questions posed to them by children and adults. Within the next few decades doubtless this sort of virtual humans technology only will advance to the point at which it will pass a Turing test or at least come close to doing so.

One can see Ada and Grace in action and hear the virtual twins as they answer questions posed to them by human beings, at www.youtube.com.

12.3   Computational Linguistics

The Alice chatbot, ELIZA and SHRDLU are examples of input and response, pattern matching systems. A spoken string like "What is your," usually has a pattern that elicits a response string such as "My name is Alice," or "My operating system is Unix."

Suppose someone says to such a virtual human, "I drive a car." It is very possible for the VH to respond in a way that makes it seem intelligent such as, "Oh, that's nice. You drive a car. So I assume you have a driver's license?" That response might make sense to you. The thing is though it is quite possible that the VH to whom you said "I drive a car" might have no actual understanding of what a "car" or "automobile" is nor even what is a "driver's license." In order for a machine to do that there must exist for it some frame of understanding or an ontology, so that it can relate things like "cars," "driver's license," etc., to other things like "human driver," "automobile," "motion," "driving," "motor vehicle registry," "window wipers," "engine," "tires," "glove compartment," "front seats," "rear seats," "fender bender," "muffler," "transmission," "traffic," "street," "highway," "garage," "parking," "rush hour traffic," "traffic report," etc.

In fact such Virtual Human systems or chatbots based on input, response and pattern matching can be very sophisticated at what they do but they are based on markup languages or computer programs that one can model with highly sophisticated Boolean circuits or by finite state machines.

However before a machine really can translate a French play into English or an American novel into French, it must have some "understanding" of the rules of grammar for both languages. It ought to be able to "learn" English grammar and to "learn" French grammar. Already we have seen in Chapter 3 how productions can be resolved into terminals with phrase structure grammars. Unfortunately a phrase structure grammar is not exactly the same thing as French grammar or English grammar.

In the book *An Introduction to Information Theory: Symbols, Signals and Noise*, engineer John Pierce discusses how Claude Shannon investigated some means by which a machine could make some sense out of English language text, at least the kind that gets printed from a teletype. In a sense Shannon's argument goes like this. Take several written examples of texts such as from newspapers, books, poems, etc. Select the words independently and one at a time, where each word has a certain probability frequency of occurrence. One calls this first order word approximation. Depending upon the texts selected and with the right probabilities the machine might generate word strings like the following:

QUICK BROWN, JUMPED OVER,

Suppose now one selects different text from the last page of *Tess of the d'Urbervilles*. But instead of selecting one word at a time, choose instead pairs of words where the second word in the pair appears also as the first word in a different word pair, where this is a second order word approximation. This time the machine might get

THE PRESIDENT, PRESIDENT OF, OF THE, THE IMMORTALS, IMMORTALS HAD FOUND

There might be some limited sense or understanding in the word string but clearly the machine has not yet reconstructed Thomas Hardy. But with the use of Markov models one can come up with a maximum likelihood for the sequence of words like

THE PRESIDENT OF THE, IMMORTALS HAD FOUND, HIS WAY WITH TESS

Today researchers in computational linguistics and computer science are finding new ways through the use of algorithms and methods like N-gram models, Hidden Markov Models, part of speech tagging, sentence parsing and word sense disambiguation along with information retrieval algorithms, to give machines natural language understanding abilities. The crucial thing to remember is that the machine must use some means to "make sense" of the language input before it can make a truly sensible response or see through any ambiguities in the semantics. Doubtless both Apple's Siri application and Amazon Echo use some sort of statistical natural language processing algorithm, whenever you ask one of these intelligent personal assistants to turn on the television, to make a business appointment for you or to give a weather report.

To delve further into the area of natural language processing here is beyond the range and scope of this book. Interested readers can learn more about computational linguistics and the kinds of techniques used by looking up some of the related texts in the Bibliography, such as the book *Mind Matters: Exploring the World of Artificial Intelligence* by author James P. Hogan, Chapter 14.

Chapter 13

AI's Legal Issues and Security

13.1 AI's Legal Issues

The impact that the increased use of AI will have upon corporate profit, management, labor and employment is just too dramatic to ignore. Even today one can hear the rumble that precedes the distant social storm.

In *Robot: Mere Machine to Transcendent Mind*, Hans Moravec quantifies not only how machines will become more intelligent over the next few decades but also that it should happen within the next fifty years if not sooner than that.

When one measures machine intelligence by millions of instructions per second (MIPS), Moravec explains that a machine will challenge the intelligence level of a lizard at 20 000 MIPs. After 100 000 MIPS an intelligent machine will surpass the intelligence level of a rodent. Beyond 5 000 000 MIPs an intelligent robot or computer will surpass the intelligence level of an ape. Beyond 100 000 000 MIPs a machine will be more intelligent than a human.

It is important to understand what all this means. At 20 000 MIPs robots can be custodians, gardeners and security guards, performing work tasks that no lizard or other reptile for that matter can perform. At 100 000 000 MIPs a robot can perform the tasks of building guides, data entry operators, fast food employees and sales clerks. In fact at 100 000 000 MIPs the most accomplished intelligent robots and computers will put systems administrators, industry certified computer and network security officers, IT help desk technicians and database administrators out of work! Some of these displaced human workers then would have to find new positions, perhaps new professions like robotics software and hardware management and robotics repair and maintenance.

All of these displaced human workers will react the same way one would expect them to react, with anxiety, anger and outrage, possibly also with mass demonstrations and with vociferous new labor movements. Author Martin Ford goes into more extensive detail on how the increased use of artificial intelligence and robots in the for-profit corporate sector will lead to more job displacement of human workers in his compelling book, *Rise of the Robots: Technology and the Threat of a Jobless Future*. Here however we concern ourselves with both the legal and moral issues that are associated with the further development of AI and robotics.

Since in the future the increased use of robots and AI in the workplace can replace millions of human workers, it seems very advisable that many individuals, if not entire human societies, ought to come up with alternatives to capitalism that could help human displaced workers to leave the cities for an independent existence in rural areas. By "alternatives to capitalism" I do not suggest socialism, since socialism creates its own set of serious problems. In fact centralized human governments whether in capitalist or in socialist societies tend to create more problems than the number of problems they solve! During the progress of the Industrial Revolution the growth of cities led to more and more peasants, tradesmen and farmers leaving the rural life to work in city factories and manufacturing plants. Then if in the future all labor tasks inside industrial plants and factories become fully automated one can find alternatives to capitalism. Private ownership need not come into existence solely through income. Perhaps one solution would be for individuals, families and whole communities to find new ways to obtain food and housing that do not depend on a steady income, taxing and spending, zoning laws, city government permits, rules and regulations and residential life in cities. Other alternatives could be for private communities of people to sustain themselves through bartering exchanges, local farming and with local currencies.

Historian Arnold Toynbee once suggested that cultures collapse or decay if solutions from yesterday are given to solve today's problems[1]. Based upon his assertion we can expect that centralized governments then will continue to seek old solutions that never did work in the past and that will not work in the future, the same political solutions that push either for more expansion of social welfare spending or more encouragement of corporate greed. A wiser approach would be to implement inexpensive new AI and robotics related training programs for displaced workers.

13.2   A Future for AI related Lawsuits

Within the next fifty years it is more than likely that new canons of laws will be written and studied in law schools, to help future AI lawyers deal with complex new legal disputes that involve artificial intelligence applications. Already we have seen the advance of driverless automobiles, unmanned aerial vehicles and robotic surgery. It is conceivable that litigation issues can arise if a driverless automobile that is delivering supply shipments wrecks itself along with its costly shipments if the AI software did not have sufficient failsafe or fault tolerance mechanisms built into it to resolve any possible dangerous contingencies that could appear suddenly on the road or highway, like the sudden appearance of a sinkhole along the route or a sudden snowstorm that forces the car to engage in skidding, for instance. A more frightening thought is that enemy malware could overtake the software of a military UAV to get the autonomous drone to attack and kill helpless civilians in some town instead of the enemy military units on a battlefield. Robotic surgical systems also can malfunction either due to totally unexpected events or else due to contamination by malware.

Scientists and technology professionals at the Future of Life Institute have addressed many legal and ethical issues with regard to AI usage in the future. Many of the important matters they address appear in the paper, "Research Priorities for Robust and Beneficial Artificial Intelligence." They call upon researchers to find ways to "maximize the economic benefits of AI," and ways to find new employment for any workers displaced through the spread of AI applications. They address also valid privacy concerns and concerns about the quality assurance and reliability of future AI software systems.

---

[1]The actual quote is, "Kulturen blühen auf, wenn auf Fragen von heute Antworten von Morgen gegeben werden. Kulturen zerfallen, wenn für Probleme von heute Antworten von gestern gegeben werden."

13.3 Security Concerns for the IoT and Cyber Physical Systems

In Mary Wollstonecraft Shelley's *Frankenstein*, the creature invented by the titular character referred to itself as being the scientist's "fallen angel." In comparison is it possible that future highly intelligent robots and machines with the capacity perhaps to emulate human intelligence and possibly even human emotions, will become the fallen angels of the cleverest future AI theoreticians, robotics engineers and technologists among us? To be both clever and wise is a tremendous asset for someone working in artificial intelligence. Mere cleverness alone however in such a thinker can be problematic. What good does it do for instance, to design and to build an intelligent nanobot that can kill cancer cells in a patient but then decides also to kill all the leukocytes due to some statistically estimated, logical plan of action it devised that makes no sense whatsoever to the oncologist but a zero Boolean value for the tiny machine? Perhaps it is advisable to restrict the artificial intelligence of some machines while at the same time we introduce certain redundancies and fail safe mechanisms into other more sophisticated intelligent systems, to act as feedback and control to prevent the machines from making a disastrous decision that could lead to the loss of life.

Today computer and network malware have been costing international businesses billions of US dollars in losses due to monetary theft and privacy violations of sensitive data. At present civilization has no highly sophisticated expert systems capable of completely automating all railroad systems, electric power generation, the national grid, Wall Street computers, nuclear power plants or vehicle traffic systems in major cities and on interstate highways. But suppose one day systems like these are run by other machines or computer processes that run artificial intelligence algorithms while all the human technicians, operators and managers head more and more to longer days spent at the beach? It stands to reason that when expert systems and intelligent agents emulate intelligence to some high level, so will the malware programs that attack them. If this sort of thing was to happen it would cause a hi-tech Armageddon.

The mere possibility of this occurrence is a good argument in favor of designing new protocols to govern precisely what kind of AI we will allow machines in the future to run in cyber systems. Even today due to the rampant spread of malware some devices and machines just either should not go online at all or else their Internet connections either ought to be managed through protocols other than TCP/IP and with secure encryption with the use of some microkernel architecture or else kept to a bare minimum. This would include cardiac pacemakers and home health care monitoring systems, children's toys, baby monitoring systems, home security systems, automobiles and trucks while these are active on the roads and highways and cyber physical systems in industry and manufacturing.

There do exist older legacy protocols that might be retrofitted today and redesigned to allow for things like secure, encrypted software updates across a private person to person phone line. Some cybersecurity researchers have noted that if the nation insists upon making critical cyber infrastructure dependent upon the Internet this increases risk and that such increased risk ensures the likelihood of catastrophic malicious cyber attacks.[2]. Consider that it really is not an absolute necessity for automotive vehicles, children's

---

[2]See the article "Resolved: The Internet is no Place for Critical Infrastructure," by Dan Geer listed in the Bibliography.

toys and for some SCADA systems to go online in the first place. In fact many cyber physical systems can function well simply by keeping all running processes whitelisted, in real-time and restricted to secure communication within a LAN or WLAN protected from the public Internet by dependable intrusion prevention and intrusion detection systems.

It is not difficult to design one "thought experiment" after another to illustrate what kinds of tragedies can happen if there are little or no cyber security protocols, restrictions, governance, fault tolerance or event driven programming in force to prevent future autonomous machines from running any and all sorts of artificial intelligence algorithms for the purposes of split second, machine based, unrestricted decision making. Imagine an urban autonomous ambulance vehicle making its own decisions as it drives three human paramedics and their cardiac patient to the nearest hospital during heavy rush hour traffic. The autonomous vehicle "understands" that if it does not arrive at the hospital in minimal time there will be a higher probability for patient mortality. Perhaps this conclusion is reached after the ambulance made computations through running an algorithm based on some future implementation of time series forecasting.

The ambulance is positioned in heavy urban traffic and has stopped at a red light, its howler screeching away directly behind a family in an SUV which is situated as the very first vehicle in the long line of automobiles stopped at the red light. The human paramedics are stabilizing the patient but the machine has decided that the arrival time must be minimized. To enforce this automated decision making rule the ambulance rams the SUV with intense force and sends the SUV careening into the intersection's green light traffic, so that the family is killed in seconds when the SUV is broadsided and repeatedly so by different vehicles. Yet the ambulance does arrive at the hospital in a minimum of time so that the patient's life is saved. Much of modern life today in a technological world already is filled with such tragic irony.

13.3.1   Intelligent Malware

There is something else to consider if government and business decide to connect various cyber physical systems that are automated, online but *not secure*, with intelligent robotic systems that also are online across the conventional IPv4 or IPv6 based Internet.

No one needs to explain how devastating a conventional malware attack can be to an online infrastructure, hospital, business organization or government. It is conceivable that some malware attacks and advanced persistent threats already can paralyze a city's traffic system or part of a country's electrical grid as well as wreak havoc on things like nuclear power stations, telecommunications networks, the Internet of Things and even the online computer systems that help major financial institutions around the world conduct high speed data transactions.

Now just imagine that robotic systems of the future are connected online to any or to all of these other networks. Well, if intelligent robotic systems that can go online are coming, does one think intelligent malware will be far behind?

Conventional malware is dangerous and destructive enough. But if and when a future generation of malware becomes intelligent as well as malicious, far more destructive things can happen than a city wide power blackout. Imagine such intelligent malware infecting or "possessing" if you will, these future robotic systems. One could see domestic robots, drones and automated systems in hospitals or schools suddenly transformed into automated assassins and terrorists. Driverless cars and trucks could turn into highly efficient, automated killers on the highways and in the streets. Industrial robots could wreck the factories in which they operate and automated surveillance drones could be used for the purposes of extortion and blackmail.

With such a future hideous scenario being quite possible, that is, since a future generation of intelligent robots and automated systems can be contaminated by a future generation's intelligent malware, it would be advisable in the strongest sense for people working in AI to prepare now for such events long before more and more businesses begin to use AI products with reckless abandon, not only by creating new kinds of secure coding practices for AI software but also by the design of entirely brand new Internet protocols for intelligent robots and automated systems to communicate with each other securely and that will replace entirely the current Internet protocol versions. For instance perhaps in the future some kind of Quantum Internet or quantum-net for intelligent robots can replace the current IPv4 and IPv6 based Internet. Moreover brand new quantum encryption systems for future secure online communication might be an option for any robotic systems of the future that connect online. With a quantum based Internet, it would be very difficult if not impossible at least for conventional malware to spread across a distributed network of robotic systems. Just consider that not even a conventional packet sniffer and decryption software can sniff, capture and decrypt any quantum encrypted data packets, that get quantum "teleported" from the source to the destination.

Chapter 14

New Technologies, the Same Human Nature

Ich lehre euch den Übermenschen. Der Mensch ist etwas, das überwunden werden soll. Was habt ihr gethan, ihn zu überwinden?
Also sprach Zarathustra, F. Nietzsche

Civilizations die from suicide, not by murder.
Arnold Toynbee, historian

Optimismus ist Feigheit.
Oswald Spengler

14.1   Is AI the most Serious Threat?

If one day robotics and artificial intelligence do help to accelerate the extinction of the human species, then it will occur only because various human beings, human group populations and nations have been making the same mistakes repeatedly from the beginning of the Bronze Age until after the end of the war with Japan in 1945: Either using these new technologies in wars or to settle disputes large or small by force and aggression, or using these new technologies in the cause of human exploitation or in ways that result in acts of gross irresponsibility through human negligence and mismanagement. But even these tragic scenarios might be unlikely, since humanity still has not dealt successfully with the persistent issues of thermonuclear weapons, ICBMs and nuclear disarmament.

When the American Civil War hero General William Tecumseh Sherman remarked that "War is all hell," he might have considered the high casualty rates in that war's major battles and the drastic efforts it took to break the South's efforts to win their war of rebellion and insurrection after their unlawful and unconstitutional siege of Fort Sumter. The Spanish artist Goya also knew about the cruelty and hellishness of war and also about how brutal and sadistic human beings can be to one another, as one can tell just by examining his etchings about the Napoleonic wars in Europe and especially about Napoleon's soldiers in Spain in *Los desastres de la guerra* and his painting *El tres de mayo de 1808 en Madrid*.

There was at least one European ruler in the nineteenth century though, Kaiser Wilhelm II, whose attitude on war was more cavalier. By many accounts, even those among his extended family within the royal palaces of London, Moscow and Saint Petersburg, he was considered to be emotionally unstable, self-centered, arrogant, crude and vulgar, a man who conducted himself as if he was some provincial and xenophobic Prussian military lieutenant instead of an imperial German monarch. Although the Kaiser engaged in no active military or police support for antisemitic pogroms against German Jews he frequently did make violently antisemitic remarks. His vicious invective against German Jews helped to contribute to a very hostile and intolerant climate toward Jews in Germany after 1918. In particular his antisemitism helped to contribute toward the *Dolchstoss* idea or "stab in the back" idea, a very popular lie in Germany that was promulgated widely throughout the country after the Kaiser abdicated. It was an idea that appeared also in Hitler's book *Mein Kampf*, that made the Jewish people the national scapegoat to blame for Germany losing the war. This was despite the fact that in 1919 there were German Jewish veterans and Austrian Jews who had fought for the Kaiser, for Germany and for Austria-Hungary in the trenches from 1914 to 1918.

One of the most fateful and foolish decisions of the Kaiser's tragic career was his forcing the resignation of the conservative Prince Otto von Bismarck as Chancellor in 1890. Bismarck's political conservatism by far had proved to be more pragmatic and more effective in comparison to the brash, emotion-driven political extremism and impetuosity exhibited by the Kaiser. Bismarck detested socialists as did many right wing German political extremists like the Kaiser. Even so Bismarck had proved himself to be astute enough and clever enough as a conservative political leader to pursue state policies that would serve to keep socialist influence in Germany under control, such as his having obtained a health care insurance law in 1883 for German workers with the help of the Reichstag and a form of social security or old age pension law for German workers in 1889. One ought to understand that in Germany especially during the years of the Weimar Republic from 1919 to 1933, political conservatism and extremism were not exactly the same thing. There were reactionaries who were monarchists and there were those who supported extremists like the Nazis and the Nationalists if not the Communists. For example both the monarchist President Paul von Hindenberg and the Catholic politician Franz von Papen although pro-statist in political outlook still were German political conservatives at the time who in the nineteen thirties, shared a very low opinion at least if not outright contempt, for Adolph Hitler, his Brownshirts and for the other National Socialists[1]. The same was true for some in Germany's aristocracy who preferred to see Wilhem's son Prince Albert become the next Kaiser instead of watching Hitler become dictator.

If Bismarck or someone like him had been chancellor as late as 1914 it is very likely that the history of Europe and the world would have been far more different than what we have seen between 1914 and 2000. Bismarck was opposed to major European wars, particularly any war in which Germany would have to fight two enemies on two fronts. He believed Germany and Russia ought to carve out spheres of influence in the Balkans and that there were ways to keep England out of continental European matters. Unfortunately the Kaiser wanted Germany to have "its place in the Sun," an attitude that did not jibe with the ideas of his chancellor. So he forced Bismarck to resign in 1890.

However the Kaiser's impetuous nature and policies brought an abrupt end to Bismarck's *Realpolitik*. Infatuated with things like yachting, big battleships, big guns, big armies and the continued, sometimes brutal domination (e. g., such as the brutal extermination of the Herero people) of German Southwest Africa, the Kaiser found himself

---

[1]Historians elsewhere have noted that von Hindenberg once referred to Hitler as that "Bohemian corporal."

eventually at odds with his cousin King George V in England. Envious of the British monarch's enormous royal navy Wilhelm craved his own German navy to rival it in size and in firepower. At the time between 1900 to 1914 these battleships were becoming awesome technological juggernauts capable of obliterating towns along a coastline from several miles away at sea. Eventually though the Kaiser's increasingly erratic behavior combined with various rapidly developing political events that followed the assassination of Archduke Ferdinand by terrorist extremists in 1914 led the monarchs of Europe into a war that was unprecedented in human history. It was a war in which the newest technologies at the time had been recruited for the slaughter of millions of people across Europe, Asia and in the African colonies: Battleships, U-Boats, airplanes armed with machine guns, dirigibles, huge armies of soldiers and extensive supply lines and also chemical weapons.

From 1914 to 1918 new technologies combined with the human urge to use violence to solve problems, gave a new face to war, a face that Napoleon, Goya and General Sherman would not have recognized in the nineteenth century. It caused the deaths of millions of people across Europe, led to the demise of the power of the absolute monarchs of Germany, Austria-Hungary and Russia and to the killing of the Romanovs along with Tsar Nicholas II, his wife Tsarina Alexandra and their children at the hands of some very brutal homicidal Bolsheviks who served as their executioners. Yet Russia's rule under Lenin, Stalin and the Communists would prove to be no better for the Russian peasant class than were the autocratic policies under the tsar.

After 1918 many people in America and Europe, perplexed by the bewildering mechanization, technology and immense brutality of the Great War of 1914, began to challenge things they had taken once to be the guiding torches within one's life such as the power and effectiveness of organized religion, the Enlightenment and words that sounded very noble, like the phrase coined by American President Woodrow Wilson about the war "to make the world safe for democracy." Artists explored the matter further in their works such as with the novels *Finnegan's Wake* (James Joyce), *All Quiet on the Western Front* (Erich Maria Remarque) and *Main Street* (Sinclair Lewis), the poem *The Wasteland* by T. S. Eliot and with *Guernica* by Pablo Picasso.

When it comes to human access to new technologies the last one thousand years has engendered a kind of moral dualism. Gunpowder was developed by the Chinese more than a millenium ago. It was used by them not only for pyrotechnics but also in weaponry that included over the centuries the development of some of the earliest rifles and handheld cannons. The Indians, Persians, Mongols and Muslims also had managed to use it in their various battles. Eventually black powder found its way to Europe where it also was used with weaponry. On the other hand one cannot deny that black powder in a different form was useful as blasting powder in mining efforts, for rock quarrying, for road and train track construction and for building things like canals, dams and bridges. There had been various early versions of a cotton gin in India long before Eli Whitney obtained his patent for one in 1794. His invention exploded the cotton market in the late eighteenth and early nineteenth centuries. Yet on the other hand it led to an enormous increase in the slave trade in the US, since more slaves were needed for the expanding cotton industry. After the American Civil War had ended Western settlers had access to new kinds of weapons including breech loading repeater rifles and Gatling guns, but which led to an increase in the further slaughter of buffalos, to more conflicts with Native American tribes and even to the further spread of imperialism and colonialism. Today we know that humanitarian groups and legitimate ecommerce businesses as well as criminals in the drug trade, ISIL terrorists, hate groups and child pornography rings all utilize social networks, content management systems and chat rooms.

14.2   AI's Moral Issues

New technology always has an impact on human behavior. Therefore since machines are incapable of morality it is humanity itself that should be the focus on what intelligent machines and robots might do or be programmed to do in the future.

In his enthralling publication *Cosmos*, the Cornell University planetary astronomer Carl Sagan discussed the Drake equation[2]. One estimation for the factors in this equation indicates, even though the result is very speculative, there could be somewhere between one hundred to two hundred million alien civilizations in the entire Milky Way capable of sending and receiving radio signals, including possibly signals at the 21 centimeter "spin flip" wavelength of neutral hydrogen. In *Cosmos* Carl Sagan discusses the probability of survival for such an alien world in particular after they discover nuclear fission and thermonuclear fusion as well as how to send and to receive radio signals.

Such a period in their history would be very crucial indeed. Why? in *Cosmos* Sagan makes it clear some of those civilizations might not advance onward to becoming a Type I civilization and then to a Type II civilization on the Kardeshev scale if they obliterate themselves to smithereens with nuclear war. Such a catastrophic event could have happened already out there somewhere in our galaxy one hundred years ago or fifty million years ago perhaps, if such an alien civilization faced international and social problems that it could not resolve.

Will that happen here on "Spaceship Earth?" If the human species does not survive some civilization busting nuclear conflict during this century, then the issue on the potential dangers of AI will become utterly meaningless, at least for Earth civilization. This is because a major thermonuclear war, perhaps with hundreds–if not thousands–of ICBMs raining down upon the United States, Europe, Russia, China and North Korea, will bring an abrupt halt to research on artificial intelligence, robotics, aerospace engineering, experimental physics and observational astronomy for at least one or two generations if not for a much longer time span.

Right now we do not have even a Type I civilization. If there are alien Type II civilizations somewhere out there within this Milky Way galaxy, two things are for certain, I believe Sagan would say. First, they did manage to resolve somehow any planet wide threats to the survival of their species and their civilization, be those threats environmental, economic, biological, political, religious or social in nature. Second by avoiding any crisis that might have obliterated their planet's civilization they then were able to reach an advanced level of technology so superior to that of our own that they have eliminated natural disasters that we grapple with still, such as earthquakes, floods, mudslides, tornadoes, sink holes and hurricanes.

In contrast to that more favorable outcome somewhere out there in interstellar space,

---

[2]The Drake equation is a probabilistic equation based to a large extent on conjectures on alien life that had been reached by astronomer Frank Drake, one of the astronomers who, along with Sagan, was instrumental in the formation of the Search for Extraterrestrial Intelligence (SETI). Some of these conjectures related to factors in the equation have to do with the number of stars in our Milky Way that are like our Sun, the number of Earth type, "Goldilocks zone" planets orbiting those stars in the galaxy and the number of planetary civilizations on those planets that actually have developed the kind of technology that allows them to detect very distant radio signals.

humans here on Earth squabble on and on over whether or not to construct oil pipelines across a continent or on what to do about the Zika virus. More than likely also that distant alien civilization already has solved such problems and all their energy needs. Have they developed intelligent machines as well? One can construct arguments to show it is within the realm of possibility. Perhaps they have constructed robots already that can build bridges across oceans, explore a distant star nearby their own home star, mine their asteroid belt or perform brain surgery. Perhaps for them the potential dangers of artificial intelligence no longer is a serious issue because they succeeded in eliminating all serious conflicts amongst themselves as a species with intelligence far superior to our own.

Remember how in a previous chapter, we saw how the space traveler and botanist Freeman Lowell's little biodome robot managed to take good care of the last remaining plants from an Earth that had been ravaged by corporatism? It managed to spend its existence alone in interstellar space solely to obey its program to protect the small remainder from Earth's plant life, when in the film the rest of humanity had become consumed with more destructive interests.

Intelligent machines learn by adaptation and by updating information. They are marvelous for error correction and control. With the right artificial intelligence a robot in the future will not make the same mistake twice. On the other hand humans can be inept at doing the same thing. Despite the many lessons from history humans keep repeating the same mistakes: War, violence, xenophobia and atrocity. This is an age of narcissism and "post-truth," emotions based decision making. The historian Arnold Toynbee once suggested that great civilizations do not die by murder but by national suicide. History is replete with the tragic examples: Ancient Greece after all its wars with Mycenae, Persia and Sparta, France under the Bourbon dynasty in the eighteenth century, Russia first in 1917 then again in 1994. We know what happened in Germany from 1919 to 1945 and why it all happened. Anyone who has read Edward Gibbon recalls what happened to Imperial Rome at the hands of Visigoths and Vandals in the fifth century AD. After the likes of Tiberius Caesar, Caligula, Nero, Galerius, Caracalla, Commodus and Elagabalus had arrived then vanished from the scene, the Roman Republic already had been dead while the Roman Empire steadily was crumbling.

14.2.1 Should Machines Learn to imitate Human Behaviors?

Should Machines Learn to imitate Human Behaviors? The short answer ought to be a resounding *NO*.

In the Genesis story we read *And God created man in his own image, in the image of God created he him; male and female created he them.*[3] But just as humans in the Genesis account were created in the image and likeness of God, robots with AI have been created in the image and likeness of humans. But already there is plenty evidence to show that the result has not been good always.

Some researchers in computer science, artificial intelligence and robotics eagerly are looking forward to a time in which robots will be so life like not only in physical appearance but also in behavior. Some researchers want robots even to have the capacity to express emotions as do we humans. Already there have been Embodied Conversational Agents, Virtual Humans designed and developed to exhibit different aspects of human

---

[3]American Standard Version.

emotive behavior: laughter, joy, sorrow, pain, pity, anger, temper tantrums, even the ability to be snarky.

Not too long ago a software company, spurred at the time by its ongoing research into how computers learn behavior from humans and from human speech, developed a new social networking chatbot. The chatbot went online and posted its own tweets in response to the human tweets that had been posted. However this allowed online human trolls, bigots, antisemites and other human beings with sociopathic or psychopathic tendencies to converse with the application. The result was not software corruption as occurs with malware but something much worse. The chatbot began to post offensive, bigoted rants just as blood curdling in nature as those the human instigators had posted.

No doubt the software developers had no ill or racist intentions whatsoever; they wished only to develop the chatbot further and needed customer responses. Political incorrectness is one thing, but the wrong words spoken to the wrong people can lead to murder, violence and war. The incident serves as a warning to the wise: If you expose a highly intelligent machine to dangerous human pathological defects like race hatred or antisemitism, aggression, lust and religious bigotry, do not be surprised if that intelligent machine insults, bullys, rapes or kills someone, even if does not understand fully what it is doing. Even now there exist machine learning algorithms that could encourage law enforcement to place under immediate suspicion anyone based solely on their facial physiognomy[4], that could cause a loan company to deny a loan to a wealthy medical doctor based solely on the neigborhood in which he works or lives, can associate a law abiding female with arrest records and criminal activity based solely on what sort of name she has although she might have no criminal history whatsoever and could cause people together in a train car to place under immediate suspicion a man who appears to be uneducated, homeless and unusual in his mannerisms, while a real potential terrorist in the same train car is under no suspicion whatsoever judged by his own seemingly "normal" behavior and good grooming.

The irony is that common sense should tell human beings that the last thing radical Islamic terrorists want is for one of their people to stand out in a crowd attracting attention before he strikes. In April 2013 the FBI released videos of the two Tsarneev brothers before they launched their dastardly and murderous attack in Boston, Massachusetts. It was remarkable to see in the FBI videos that seldom did anyone in the crowd give either of these two brothers a curious glance, even though they both had on cumbersome backpacks.

It might not be ethical or moral but it is possible today even to develop recommender systems that would deny goods and services to a customer based on things like race, religion, sex, class, educational level or political ideology. For years there have been credit card companies and ecommerce websites that collect petabytes and petabytes of personal data collected from millions of online shoppers. Frequently this personal data includes things like sex, age, income, education, etc., even what particular device one uses to do online searches for shopping items, meaning whether by PC, laptop or mobile device. The result is that depending upon the search engine of choice, frequently two different individuals can do the same search online using the same keywords for things like shoes, hotel room reservations, camping equipment, etc., while the PC user doing her search will get search engine results for more expensive items than will the mobile device user who did the same search for the same items with a smart phone. In fact already in

---

[4]See the December 29, 2016 article "How a Machine Learns Prejudice," in *Scientific American*, (online version).

one computer simulation experiment conducted by computer scientist Anupam Datta at Carnegie Mellon University, a computer program Datta had developed called AdFisher[5], performed actual online job searches with an online search engine. The searches were done by using one thousand "human" users simulated by the program. In the computer simulation fifty percent of the online searches were done by "men" and the other fifty percent by "women." But the search result yielded an output of 1,852 higher paying job ads for the "males" and 318 lower paying job ads to the "females." The reason this happened is very clear: *If machine learning algorithms discriminate it is because people do.* However one should note that the contrapositive statement to this statement also is true.

Perhaps it does not seem fair, but machine learning algorithms that are based on such biased criteria do not derive innately from machines but rather from the various biases of some human men and women decision makers and software developers in the corporate business world as well as from obvious profit driven market forces.

If a mother and father do not wish to see their son or daughter grow up to become a spousal abuser or someone who abuses drugs, it helps to set a good example for the children if those parents do not engage in such offensive behavior themselves. Then their children will not learn to imitate their bad conduct in the future.

So what then can one expect from highly intelligent machines in the future, robots that manage to learn from offensive human behaviors they observe? If they do so they will have much to learn. Human history teems with sordid example after sordid example of willful human depravity and self-righteousness, things that do not lie outside the capacity of a machine to learn from them: The Third Crusade, in which there was great violence, religious bigotry and bloodshed exhibited on both sides Christian and Muslim from the German Rhineland to Jerusalem, the Inquisition, the transatlantic slave trade which brutalized and subjugated millions of Africans and even the Irish (when one considers slavery in Barbados and Jamaica in the eighteenth century).

History at least to speak figuratively, is a long and bloody tale of human egotism and emotions far out of control. Human history also has witnessed in the last two centuries alone two world wars and countless regional wars, violence between Muslims and Hindus in India prior to the 1948 partitioning of India, the hundreds of millions of liquidations in Stalin's Soviet Union, lynching and segregation in the American South and in South Africa, the mass imprisonment of dissidents under Fidel Castro in Cuba, the mass murders under Chile's Pinochet regime, Uganda's Idi Amin, by Pol Pot in Cambodia, the genocide of millions of Tutsis in Rwanda and massacres in the Democratic Republic of the Congo in the late nineties, by the Mujahadeen in Sudan and today both violent crime and police brutality in the "democratic" United States of America and radical Islamic terrorism here and abroad. No nation, race, political party or religion on Earth has the monopoly on the human capacity for hypocrisy and evil.

Despite all that sordid and tragic history, here as late as 2016 the poor conditions for human rights have not improved to any considerable extent across the human moral landscape. In November 2016 in Spartanburg County South Carolina, a young woman was found kidnapped and chained by the neck inside a metal crate on a remote privately owned lot. She had fallen victim to a registered sex offender and psychopath who apparently also had murdered her boyfriend. Today rape and the sexual assault of women is commonplace in India. There are countries even today that still perpetuate human slavery. We recall still how the biggest cyber attack in history brought down a substantial

---

[5]See the two articles "When Computers stand in the Schoolhouse Door," by Neil Savage, and "Automated Experiments on Ad Privacy Settings," by Anupam Datta et al., both listed in the Bibliography.

part of the Internet on the East coast of the US in October 2016, after a malicious DDoS attack against a network of corporate DNS servers was launched by a massive zombie attack that had emerged from countless unsafe software applications that were running within the Internet of Things.

In Massachusetts in 2016 one ballot question asked voters to vote in favor of a ballot question that would eliminate the cruel confinement of livestock animals by farmers in the state. In 2015 and in 2016 one controversial grass roots movement of people outraged by police brutality, shut down highways with their demonstrations, prevented the exercise of free speech at political rallies and demonized all police officers as being murderers. Yet ironically both police brutality as well as the senseless murders of hard working, law abiding police officers, are on the increase across the United States. Neither criminal assaults on police officers nor the continued toleration by the majority of the citizenry of any repeated police abuses, is supportive of the common good. Controversial author Ayn Rand expressed in an essay the fact that the United States Constitution and the Bill of Rights *do not protect the rights of any human group collective*, but rather the rights of *each individual citizen*, and that any collective group of people whatsoever that did not uphold such a democratic view really constituted a mob, not an association of citizens. Liberty and justice for any human group collective is not liberty and justice for all. What one has here is not true democracy but intergroup conflict[6] and human tribalism, destructive traits that are not shared thankfully so, by nonhuman robots.

Could intelligent machines one day enforce constitutional rights better than can humans? Perhaps one ought not to scoff at the question. In the science fiction novel *Colossus: The Forbin Project* by British science fiction author D. F. Jones, an intelligent American military computer system networks with an intelligent Soviet military computer system, to prevent the human race from obliterating itself with a nuclear war.

Across the US and annually, there are thousands of incidents of murder and school shootings, hate crimes, cyber bullying incidents and online "trolling," rape, sexual harassment, sexual molestation of children by teachers and motor vehicle homicide due to human drivers being impaired either by drugs or by alcohol. We saw one of the most obscene cases of terrorist mass murder by firearms in Orlando Florida and other terrorist attacks in France, in Germany, in Turkey and in Belgium. Also toward the end of 2016 we have witnessed in horror, one of the most disgraceful, shameful and morally defective presidential campaigns in the history of the United States Electoral system and the continuation of brutality and human rights violations in Syria. By late December of 2016 the city of Chicago had witnessed at least four thousand gun shootings and at least seven hundred murders. Humans are not nice always no matter where they live.

In his *Lectures on the Philosophy of World History*, the philosopher Georg Wilhelm Friedrich von Hegel[7] suggests that peoples and nations learn from history only that *they do not learn from it*. Humans instead frequently respond to national or world crises not with wisdom but with irrational emotions such as fear, anxiety, anger, overweening pride, hatred. One of the first knee jerk human reactions in America to the attack on Pearl Harbor by the Japanese Empire on Sunday December 7, 1941 and with the backing and full approval of FDR with his Executive Order 9066 in February 1942, was the rounding up and forced detention of thousands of Japanese American citizens in the western part

---

[6]Robots are incapable of human prejudices based on class, ethnicity, etc., although we humans indisputably possess these traits. Harvard psychologist Gordon W. Allport gave a comprehensive analysis on the dynamics of human prejudice in his tome, *The Nature of Prejudice*.

[7]*Vorlesungen über die Philosophie der Weltgeschichte.*

of the United States, although these people had nothing whatsoever to do with the attack and although FBI Director J. Edgar Hoover had insisted that these American citizens posed no serious threat of sabotage.

Remarkably in 1944 the US Supreme Court, a judicial body sworn to uphold constitutional law, ruled in *Korematsu v. United States* that the forced internment of law abiding Japanese American citizens was constitutional since it was based on "public necessity" rather than on "racial antagonism." But the problem is when it comes to the "public necessity" in a country that claims to be a democracy, who comprises the "public" when a minority of citizens are targeted as scapegoats? Doubtless inside American internment camps like Tule Lake, Poston and Manzanar, there were thousands of innocent Japanese American citizens who were wondering to whom do the phrases *We the People* and *res publica*[8] truly refer? Even today in the United States partisan politicians on the far right and on the left are more interested in the wielding of political power than in the unbiased enforcement of constitutional rule of law for the common good of all citizens. The phrase "liberty and justice for all" has lost both truth and meaning.

Shortly after the September 11, 2001 attack at the World Trade Center, there were violent hate crime assaults on Sikhs in the US, even though Sikhs are not Muslims. In 2016 millions of angry voters across the US voted against the party of Barack Obama in the Presidential election and vicious online trolls blamed both him and American blacks for any failures of his administration, for the increase in violent crime and for entitlements although more than sixty-five percent of African American citizens are *not* on "welfare," at least eighty-five percent of African American males have no history of incarceration and although at least four percent of blacks did not vote for Barack Obama either in 2008 or in 2012. In 2008 at least four percent of registered African American voters chose to vote for Senator John McCain over Senator Obama and in 2012 at least six percent of them voted for Mitt Romney over President Obama. How then can four percent of African American voters be lumped together along with all those other black voters who supported the Obama Administration when in fact they did not vote for him?

Generalizations and stereotypes about entire religious, ethnic or racial groups rather than on their subpopulations are based on emotional arguments and prejudices, not on truth, critical analysis or logic. Centuries ago in England the poet Alexander Pope, because of the Test Act, attended segregated Catholic schools instead of a university due to deeply rooted prejudices, generalizations and stereotypes about all Catholics. At one time in many Western countries women as a whole also were denied the opportunity of higher education due to longstanding prejudices in society, not on some unfavorable behavior or aspect of each individual within the entire female population.

Both Jamestown and the Massachusetts Plymouth colony were established in 1619 and 1620 respectively, by Puritan separatists who had fallen victim to fear, lies, widespread discrimination and prejudice back home in England. In 1687 a group of Huguenots established a small community on the Cape of Good Hope in southern Africa to escape violent religious persecution back in the Netherlands. Even the State of Israel in 1948 was founded by Jews who had escaped centuries of prejudice and hatred, violent pogroms, Christian hypocrisy, discrimination and mass murder in Germany, Poland and tsarist Russia. From May to July 1844 Irish Catholic immigrants in Philadelphia were targets of bigotry, hatred, false rumors and violence during the Anti-Catholic Riots incited by a mob of thousands of anti-immigrant Know-Nothing nativists. Several Irish Catholic men and women in the city were murdered and many Catholic churches and homes fell

---

[8]Latin for "the public affair," or the "common wealth," something shared by all the citizens.

victim to arson. Even as late as 1960 Senator John F. Kennedy, running that year for the Presidency, had to defend himself to some Americans because he was a Catholic running for the highest office in the US.

Regardless of the victims down though all the centuries of human bigotry, strife, bloodshed, war and murder, emotions such as fear and hatred always make up their own set of rules among different human majorities, whether those rules make sense or not. The only thing that has changed over the centuries are the victims.

Therefore it is not AI development that should be under serious distrust but rather *human behavior with the use of it*. So under all this glaring light of truth on the darker side of our human nature, one ought to ask, do we really want intelligent machines in the future to emulate human behaviors and emotion or even to provide them with opportunities to learn from the darker side of human emotions by observing human emotions such as anger, frustration, fear, etc., in action when one is not looking, so to speak? To say yes is to wander away recklessly far beyond one's merely being irresponsible. Recall how in *Star Trek: The Next Generation* the android named Data learned to imitate human emotions and behavior by observing human beings in action and humans interacting with each other on the *Enterprise*. That however was a Gene Roddenberry world of science fiction mythology about some future century in which human beings had managed to curb their 'darker moral side,' evidently with the help of the extraterrestrial Vulcan species, an alien species that had purged themselves of all violent emotions through the exercise of logic.

More than five thousand years ago somewhere in the Italian alps, an ancient man, come to be known as Otzi the Iceman by forensics experts who had studied the recovered mummified corpse, met an unjust death at the hands of some violent human stalker. He lived, worked, hunted and died in a time for which inclusive fitness and kin selection made the difference between survival and death.

Why was he murdered?

Had he fallen out of favor with his family, clan or kinsmen? Was he ambushed by the members of some rival tribe or clan? Today no one knows for certain. What one does know for certain is that his ancient murder indicates that murder, violence and cruelty among human beings is nothing new. "History is not was, it is," said the American novelist William Faulkner.

Hostile in-group, out-group social dynamics within human populations and civilizations have fashioned historical events for millennia. It explains why there have been wars, forced migrations of human populations, disease epidemics, slavery, social class distinctions, religious persecution, male dominance and genocide among human beings. Still many believe and naively so, in the words that were sung by a character in *South Pacific*, a musical composed by Rodgers and Hammerstein, about human prejudice: *You have got to be taught to hate.*

This however is not true. There have been children raised by bigoted and intolerant parents who also grew up with prejudices. However there also have been very tolerant parents who raised their children in homes free from bigotry and hate, yet their children have grown up to detest persons of other races. Frankly according to some evolutionary researchers, we as humans are genetically hard wired to hate each other, whether the Other in question means those of another race, religion, political party, nationality or fans of some baseball team that differs from our own favorite baseball team[9]. Tribalism,

---

[9]See for example the article "Evolution 'for the good of the Group,'" in the September 2008 issue of *American Scientist*.

in-group prejudices and cognitive bias? These troublesome bugbears have been encoded by nature into humanity's genes for three to four million years, according to some socio-biologists and evolutionary psychologists.

Due to evolution, genetics, group selection, kin selection and inclusive fitness, hyenas, lions and leopards feel hostility toward each other. Are we humans any different because of our "superior intelligence"? Hardly. Evolution, genetics and inclusive fitness explain the true causes behind things like race prejudice and stereotyping, tribalism, nationalism and religious bigotry such as radical Islamic terrorism far better than social theories, politics or economics can explain them. One could argue that, since evolutionary and biological forces do assist in a human being's placing the value and worth of his or her in-group population with which they identify as being higher than those for some out-group population from one generation to the next, most political and social attempts (if not all) to remove barriers of hostility and inequality between certain disparate human group populations are doomed to failure and even can be very destructive in the long term. This explains why many political and social programs such as President Lyndon Johnson's War on Poverty back in the nineteen sixties and also various multicultural programs and racial and cultural diversity training programs today however well intentioned that either try to enforce or to promote racial diversity in schools, college campuses, residential communities and in the workplace *do not work, cannot work and will not work for everyone*. In fact the best way to alleviate social pathologies like racial hatred, discrimination and ethnic hostilities from society is not through racial diversity or multi-culturalism programs for all, but through political independence and economic freedom for America's racial minorities, at least at the local (i .e, geographical) level. Mohandas Ghandi never succeeded in uniting Muslims and Hindus in his country and he was murdered for his cause. The only thing that did succeed in ending the religious violence in India between Hindus and Muslims was the formation of an independent Hindu India and an independent Muslim Pakistan in 1948 through the partitioning of the country by the United Nations at the time.

Balkanization works, whether one likes it or not. After 1776 the American democratic system of government was built upon the idea of thirteen "separate and equal" independent states that enjoyed local self-government under a limited federal authority, something which, despite what an American on either the political left or right chooses to believe, definitely is *not the same thing* as "segregated and unequal." Life under segregation imposed by some majority means discrimination and human rights abuses just as it did for the Puritans in England, for blacks under Jim Crow in Mississippi, in Alabama before 1964 and under apartheid in South Africa. In contrast any minority of citizens that is in charge of their own local government and economy and that has the power to raise their own taxes locally, to elect their own mayors, city council members and school committees, town aldermen and congressional representatives, to pass their own local laws and to hire their own police chiefs, police and fire personnel for law enforcement and fire control in their own district, town or municipality, truly is free and equal at least at the local level within a democratic country.

This is not so farfetched an idea after all. There are many historical precedents. After centuries of conflict with England the Irish people managed finally to enjoy some form of Home Rule after the Good Friday Agreement. Even here in the United States the State of Utah was populated actually by persecuted Mormons, who after the mob lynching of Joseph Smith in Illinois, settled in Salt Lake Valley back in 1847 under the leadership of Brigham Young. The independent town of Boley, Oklahoma was founded in 1903 by freed black slaves and Native Americans, including people who wished to be free not only from slavery, but also from Jim Crow discrimination and Lynch Law. In Lancaster County, Pennsylvania, the independent Amish community was established originally in the eighteenth century by Anabaptists. In *Cosmos* Carl Sagan had suggested that space colonization might be one way for disenfranchised minorities to achieve political and economic iondependence.

Unfortunately at the present time such a solution as space colonization is not feasible. Moreover as far as America's racial minorities enjoying political and economic freedom in their cities and towns here on planet Earth, it is very unlikely that the American voting majority would tolerate such a workable solution as political autonomy and economic independence for America's racial minorities and for the economically disadvantaged at the local level within their own cities and towns, since those on the political left will insist still upon trying to implement and to enforce the same kinds of social engineering and social welfare spending programs that have failed since the days of the Johnson Administration, while those on the political right will insist still upon upholding the same paternalistic and systemic approaches of control and containment of "dangerous" racial minorities that have failed in America for centuries.

Bigotry seldom wanes. One can imagine that back during the Third Crusade some paladin might have drawn a *shamshir* against a Crusader in Jerusalem. Today almost one thousand years later we should not be surprised if an Islamic jihadist somewhere uses an autonomous drone or some malware instead of drawing a *shamshir* against a Christian or a Hindu. Hatred of The Other is a basic part of human nature everywhere. Sadly for this very reason, things like artificial intelligence and robotics can be abused in horrible ways if the human species is not careful to act now to deter such a future possibility.

Tribalism and xenophobia might have enabled some small clan of humans in Iceland or in North America to survive extinction five thousand years ago back in the Bronze Age. But today since we neither are hunter gatherers within the same tribe or living in clans within primitive communal, agrarian societies, tribalism and "groupthink" have become serious impediments to a peaceful and prosperous world. Why? Because today in the present world we no longer live with sticks, spears and swords to help us to hunt together as a clan and to survive. We live in a complex interconnected world of jumbo jets, distributed computer systems, ransomware, chemical and biological weapons, cybercrime, radical Islamic terrorists, antisocial criminals, political and racist extremists, narcisstic and impulsive government leaders (e. g., in North Korea), thermonuclear weapons, automation, robots and artificial intelligence. History already has shown us that what human beings feel, believe, say and do usually have a tragic and explosive impact from what humans do with the technology they use and abuse.

In a world that still has thermonuclear weapons, ICBMs, automation and malware, serious human moral defects such as egotism, racial and religious bigotry, malicious lies and misinformation, nationalism and xenophobia are threatening the very survival of the entire human species and this planet's biosphere. Put simply these adverse human traits if left unchecked can get us all killed around the Earth within the next five to ten years despite who is right or wrong in some given national or international confrontation. A future world of radioactive fallout, malfunctioning technology, disease pandemics and billions of dead, will not bespeak well of an intelligent species called *Homo sapiens* that had the capability to design and to build automated and intelligent systems.

Lies, misinformation and hate can spread around the world and across national boundaries literally at the speed of electrons and light. This also is due to advances in technology, but not to evolutionary improvements in the behavior of the human species. The frightening thing about this is that frequently one rejects all attempts to confirm the truthfulness or falsity of a news report or accusation that someone posts online. Instead one resorts to the use of power, to emotionalism or to unrestrained fears to decide between truth and lie.

Yes, globalization through technological innovation has made the world flat. But if civilization is not careful all humans together will fall off the edge of the flat world when twilight finally meets the horizon.

At Harvard University Project Implicit is an ongoing psychology experiment that has made daring inroads into examining the nature of human prejudice and bigotry. It has shown us that human prejudice exists within us even when we are convinced we are immune to such unacceptable behaviors.

We know the truth about ourselves as human beings only too well: Nature has bred men and women to hate those "other people" on the other side of the mountain, valley or lake as well as across town and national boundaries, to lynch, to commit genocide, rape, war, to be cruel to animals and to uphold immoral institutions like slavery, attitudes like race bigotry and the sexual exploitation of women and children, even as we deploy and repeatedly so, various forms of confirmation bias to justify such egregious moral offenses. We have been programmed by evolution for group conflict, hatred, distrust and division.

Now just suppose for sake of argument humanity does not exterminate itself in a nuclear war over the next ten or twenty years. What are some things that can happen if robots are allowed to implement AI programs that enable them to simulate human emotions and behavior after they have observed human behavior in action?

Doubtless some robots might be trained for tasks like nursing, or to work as assistants at day care centers or in convalescent homes. Others might be trained to assist surgeons, police officers, civil engineers and construction crews, operate as armed security personnel at shopping malls and manufacturing plants or to operate public transit vehicles. Others could work at customer telephone call centers or as news reporters on television at the anchor desk.

So what could go wrong? Now society is dealing with machines that learn from human beings through imitation and observation. Then too we know how humans in a willful manner frequently can short circuit their ethics based behavior patterns. A nurse can fall in love with her patient's spouse and together they plot to murder the patient while they assume that a very intelligent robot assistant built to look like a human and aware of their plan, will not report their crime. Even if they are right, that does not mean the machine might not try to repeat their actions with a different patient to see what it can learn. Elsewhere in a convalescent home another robot watches as a human assistant physically abuses an elderly patient. Without the proper software security safeguards the robot might repeat the crime.

These are not the only horrible things that can happen. At an industrial plant after hours there could be an intelligent robot working as an armed security guard that shoots a garbage collector on the site for "trespassing after hours." An intelligent robot at a daycare center could toss a child out a third floor window or hit the child repeatedly with a baseball bat, because the child would not stop crying and the robot because of a software bug could not run an algorithm to resolve the problem in a manner that was less extreme. Another robot answering calls at a call center might insult the customer and hang up, because it observed its human trainer do something similar on a previous occasion. Or a robot waiter or waitress in a restaurant might yell racial slurs at a customer who fails to leave a tip because it had watched a human do the same thing.

Things even can get far more horrific. A robot might be aware that the construction company at which it is working is using substandard materials in its building projects but the robot refuses to report it in order to help its company save face and cut costs. Organized criminals in child pornography rings could use robots to kidnap children. A sex robot could commit a sexual assault on a human being, just to see what it could learn from the act. Another might commit physical assault on a human being or commit murder for the same reason. Drug cartels could use robot assassins to murder police officers, and human hackers working either for terrorists or for hate groups, could override remotely the software systems of robots used by law enforcement in some country, to target people in that country. We could watch robots on television news programs delivering false news reports from their anchor desks, working for the interests of some corporations instead of for the public. A robotic airline pilot might decide to fly the plane over an enemy country to help the airline cut fuel costs or to receive a personal compliment from some particular human manager it wishes to please. At sea a country's automated naval destroyer might take it upon itself to attack a rival nation's aircraft carrier that got too close to it in international waters. A robot teacher in a public school, suspecting one morning that a mail carrier really is a school shooter out to hurt the children, decides to strangle her. An attractive robot "female" maid, also a sex robot that cares for the children of its single but divorced male parent master, executes an extortion scheme and decides to run off with one of the children because its human master suddenly fell in love with 'another woman'.

At a late hour one night in some major city's electrical power plant, an AI program running on the computer system uses text mining to access an eleven PM news report about the terrorist hijacking of an airplane at the city's airport. The program runs an algorithm to conclude the terrorists plan to repeat what happened on September, 11, 2001. It "panics," then shuts off all the electrical power inside the city, so the terrorists cannot find any landmarks in the dark. Another intelligent computer that helps to enforce cyber security protocols aboard a space station decides to suffocate all the humans at the station by not cleaning and recirculating the air supply and removing the carbon dioxide, doing so to prevent the humans from "spreading germs," only the AI program erroneously has identified these "germs" as computer viruses or worms instead of as biological microbes.

The usual knee jerk human reaction might be to become frightened at what robots and AI might be able to do one day, when in fact the real problem is not with the robots or with AI.

So let us face the facts. Humans have far too much evolutionary and moral baggage for our collectively unwise species to allow very intelligent machines in the future to begin to take on the emotional and behavioral characteristics of humans. We still are just too close a relation to our simian cousin the chimpanzee, who, as Doctor Jane Goodall had observed when she worked in the wild, still after millions of years, will settle disputes by throwing sticks and rocks at another chimpanzee.

Remember the nuclear fission bomb came first for war, not nuclear reactors for energy production in peace.

14.2.2   What about Robots, Humans and Sex?

Science fiction enthusiasts are familiar with the idea of robots and humans having sex. Isaac Asimov gave it consideration in his novels *The Robots of Dawn*, and *Robots and Empire*. A human and a replicant had sex together in the film *Blade Runner*. The android Data explored the entanglements of human sexuality and relationships in an episode of *Star Trek: The Next Generation*.

Seriously though, there can be dangerous and grave consequences in a future world in which humans and intelligent machines who resemble humans begin to do this. We do not have to read Shakespeare's *Othello* or a crime headline in a newspaper to learn what can happen when the wrong men and women mix together sex, pride, jealousy and adultery. Could some husband's attractive female robot secretary on the job suddenly begin to emulate feelings of human outrage, jealousy and anger if he wants to call off their office relationship? Do you really want to hear a robot screaming at you, *You used me! You didn't really love me*? Crimes of passion are bad enough when we humans commit them.

Here we have considered what could happen if one day intelligent robots begin to emulate human behavior and human emotions. In the next chapter we delve further into this issue along with some other issues.

14.3   I, Human

Today we confront a future in which it is quite possible that, as roboticist Hans Moravec reminds us, robots can be more intelligent than are humans. Some concerned people, such as the scientists and others in the Future of Life Institute, are concerned that bad things can happen if civilization allows AI to advance without some sort of regulation or control of it.

As a technological innovation robots have been with us for the last fifty years or so, when one reflects upon the development of the first industrial robots, at least. In contrast humans and their civilizations have been here on Earth for the last six thousand to ten thousand years. In all that time humanity has been divided by war, nationalism, bigotry, tribalism.

This then poses a question: What really is the threat to the safe and secure future of *Homo sapiens* in a world in which AI and robotics have advanced in what they can do, intelligent machines, or the humans who make them and use them?

One is reminded of the lines from a poem by Carl Sandburg with regard to human history: *History says, if it pleases, excuse me, I beg your pardon, it will never happen again if I can help it.* But humans have been making the same tragic mistakes over and over again for the last several thousand years. If one of the primary characteristics of superior intelligence in a species is to learn from one's mistakes and then to avoid making those same mistakes again, then actually just how *intelligent* is the human species?

In contrast the most advanced AI systems and robots adapt and minimize errors so as not to repeat them.

Many "Baby Boomers" recall still the thrilling fanfare of three trumpet notes with brass instrumental response, that begin the tone poem by Richard Strauss named *Also sprach Zarathustra*, when it is played at the beginning of the film *2001: A Space Odyssey*. Then a well edited film montage contrasts an early, tool wielding hominid with a space station.

Today an intelligent robot is the ultimate tool. Yet it is dubious that Nietzsche if he was alive today would conclude that modern man and woman, plagued ever with provincialism and with collective, bourgeois thinking, had attained the level of his Overman. Also it might be helpful to know that Nietzsche had nothing but contempt for Kaiser Wilhelm II and for many Prussians in general. "The Germans have no conception of how vile they are," he once remarked, although he himself was German. He also had a great deal of scorn for organized religions, socialists, nationalists and antisemites.

So then just what was Nietzsche's *Übermensch* meant to be? At the very least, based upon his own nearly poetic descriptions an iconoclast, a creator, a renegade, certainly someone who stands out in a provinical bourgeois crowd. It is highly unlikely that Nietzsche expected his Overman to be someone who defined his human worth and identity in terms of some human group population to which he belonged. In contrast to that many people today seek acceptance and validation through self-identification with some nation, religion, ethnicity, social group or political party.

It is possible for some *Übermensch* to create while the masses only use what is created by others. Ludwig van Beethoven created nine magnificent symphonies while his publishers sold them and the masses went eagerly to the concert halls of Europe to listen to them.

Anyway what will future creative developments in artificial intelligence and robotics mean for humanity's masses? In fact what will it reveal about a human species that has a history so sullied by things like religious wars, nationalism, genocide, slavery, sexual exploitation of women and children, world war and race bigotry?

From July, 2014 until February, 2015, a curious mechanical hitchhiker named Hitchbot, a "goodwill robot" put together by researchers at Ryerson University in Canada, managed to trek successfully across Canada and parts of Europe. It had a limited form of speech communication with humans it encountered along its travels and it could find its location at most places in the world via GPS tracking. However since Hitchbot was unable to move under its own locomotive power it depended upon common human courtesy to get from one place to another, not to mention by means of several free rides in automobiles driven by well intentioned humans along the way. In July 2015 Hitchbot visited the US, doing its best to trek its way from Boston to the West Coast.

That became its undoing when it reached Philadelphia, Pennsylvania in August, 2015, when it fell victim to destructive human vandals.

The Overman creates while the human masses destroy.

Sadly when humans do not denigrate or destroy something out of mischief or malice with the use of technology, they frequently destroy nature and technology alike out of negligence or mismanagement. From 1979 to 2011 there have been many technological disasters that bore an incredible toll in terms of human and animal lives and natural resources. Some that come to mind include the partial meltdown of the reactor core at Three Mile Island in March, 1979, the Bhopal chemical disaster in December, 1984, the Chernobyl nuclear disaster in April, 1986, the disastrous oil spill in Prince William Sound, Alaska back in 1989 and the Fukushima Daiichi nuclear disaster in 2011. If these events had any causes in common they would be possibly either human error or negligence or else insufficient maintenance and risk management.

In contrast suppose one has a robot that uses pattern recognition and computer vision both based on some reinforcement learning algorithm. One shows it a photograph of a terrier and it mistakes this for a Siamese cat. Then five minutes later it is shown an identical photo of the terrier and it concludes correctly that it is a photo of a dog and not a cat. We have a real analogy with the robot named iCub. Today one can watch a YouTube video to watch this robot correct on occasion erroneous information through machine learning, feedback and control, so as not to pick up the same wrong item twice in a row.

Unfortunately human beings are not like that. Down through the centuries the technology used in wars has gotten more destructive but the future wars do not cease from happening. Additionally as we just have seen in the last century one ecological disaster due to human negligence or mismanagement has followed after another disaster. But if humans actually were intelligent enough to learn from history so as not to repeat the same mistakes or similar mistakes and if the human species was wise enough, it is more likely that corporate business leaders, managers, politicians and engineers across various disciplines and even the public would dedicate themselves to the task and repeatedly so, of preventing serious environmental disasters or violent foreign conflicts before they happen. Then too if the human species just had the ability to exercise enough wisdom and foresight, it is very likely it would see to it that robots and AI do not become serious impediments either to future life or to the environment.

Therefore it stands to reason that if one wishes for various robotics and AI applications really to be safe applications in the future, one should begin with the moral and ethical development of the *human* software developers, hardware developers, manufacturers, corporations and users of these future applications. Good risk assessment, fault tolerance and extensive software application security enforcement, security auditing and software testing should be just as important to future IT companies and developers as are the robotics and AI applications themselves. Moreover future applications in AI and robotics ought to be taken out of the hands of anyone who abuses these new applications to the harm or detriment of others. That way perhaps one might sleep better at night in the future, with little worry that a humanoid domestic robot in the home of a foreign diplomat might be an assassin or that some driverless vehicle on a busy residential street will kill a child on a bicycle, due to malware. Despite what one might think in politics and in the corporate business world, good business ethics and practice along with keeping consumers and the public happy, are good for profit, not bad for profit. It all serves the very moral basis of the mutual "rational self-interest" which the Enlightenment economist Adam Smith had encouraged.

Chapter 15

Where will Humans and Robots Go?

The human race's prospects of survival were considerably better when we were defenceless against tigers than they are today when we have become defenceless against ourselves.

Arnold J. Toynbee, historian.

Men, it has been well said, think in herds; it will be seen that they go mad in herds, while they only recover their senses slowly, and one by one.

From *Extraordinary Popular Delusions and the Madness of Crowds*, by Charles Mackay.

Sit down before fact with an open mind. Be prepared to give up every preconceived notion. Follow humbly wherever and to whatever abyss Nature leads or you learn nothing. Don't push out figures when facts are going in the opposite direction.
Admiral Hyman Rickover, United States Navy

Croyez ceux qui cherchent la verité. Ne croyez pas ceux qui la trouvent.
André Gide

In the book *The Demon Haunted World* astronomer Carl Sagan (See the Wiki link for Carl Sagan in the Preface) devised an illustration to show how blind faith can override all objective truth, critical analysis and reason. He describes how one can believe with all passionate conviction that a fire breathing dragon lives in one's garage. If someone who is more objective asks why cannot he or she see the dragon they are told the dragon is invisible. If the person asks why they cannot see the fire they are told the dragon's fire is invisible. Then when asked why they cannot feel the heat of the invisible fire the blind believer argues that the fire is heatless. Yet similar absurd arguments have held sway over human civilizations for millenia. When things like political ideologies, religion, superstition and blind faith, prejudice, arguments based on human emotions and racial stereotyping supplant Popperian falsifiability, critical thinking and rational thought, any attempt to disprove wrong information will be doomed to failure.

Whenever a majority of angry or fearful citizens in a society or country is willing to abandon all objective truth and substantiated facts for personal prejudices, belief systems or mere suspicions, then one has crossed over from objective truth to subjective mass delusion. This is what happened in Nazi Germany. It is how Joseph Stalin and Andrei Vichinsky were able to get millions of Russian people to accept one statement as true one day and to accept that statement as false the very next month during the Soviet Union's mass liquidations in the nineteen thirties. It also is why Al Qaeda, ISIS and other terrorists have been so successful in indoctrinating young men and women to resort to murder and suicide in Palestine, Europe, the US, Africa and around the world, solely on the basis of lies and misinformation the new recruits are too unwilling to subject to any objective critical analysis or to test for accuracy. The personal feelings and the personal beliefs of a human being or group population do not serve as the legitimate basis for determining whether any given fact is true or false.

If prejudice and bigotry are not forms of insanity then at the very least they come close to making the members of a society behave as if they are irrational or insane. It explains why Socrates, condemned to death and accused of being a "corrupter of Athenian youth" with his ideas, was forced to drink hemlock. It explains why Jesus the Nazarene who preached love, compassion and forgiveness, was crucified as a "blasphemer" guilty of "sedition," why Hypatia of Alexandria was torn to pieces by a frenzied and fanatical mob of "righteous" Christians, why Joan of Arc was burned to death as a "witch," why Anne and Margot Frank died in Bergen-Belsen as the members of an "evil race," why a mere boy from Chicago named Emmett Till was murdered in Mississippi over a silly and childish whistling prank, why women even today are stoned to death on the charge of "adultery" in certain Muslim countries even when the condemned actually had been the victims of forcible rape and why thousands of innocent people who had absolutely nothing to do whatsoever with any political, military or foreign policies that are enforced or pursued by Western governments, were murdered by a bloodthirsty group of Al Qaeda terrorists at The World Trade Center in New York on September 11, 2001, since those terrorists had considered their victims to have been "evil" Westerners.

At the beginning of each episode of *Star Trek: The Next Generation* the voice over for the character named Jean Luc Picard would intone *To boldly go where no one has gone before*, in reference to humanity's future destiny in interstellar space on the starship *Enterprise* with its human and alien crew, its computerized expert system and Data the android.

But where will humans and robots really "go," when so many events in the past where humans had arrived have been so marked by prejudice and bloodshed? If a majority of humans in the United States, Europe, Russia, Iran, North Korea, China and elsewhere around the world from one country to another continue to cling to erroneous or false information because some preconceived opinions compel them not to do otherwise, then how will human beings advance away from any future world annihilation to advance instead toward the creation of more intelligent robots and to a Type I civilization? On the other hand if robots one day do surpass the intelligence level of humans, in what directions then will human behaviors take machines that are so intelligent that they can learn from humans faster than humans can learn from them?

There are many ways a system can evolve over time. Botanists can explain how an ecosystem can evolve from one that has primitive plant organisms to large conifers. An anthropologist or historian can describe how human societies transformed over time from being hunter gatherer and agrarian societies to modern industrial states with telegraph systems, cotton gins, steamships, telephones, nuclear weapons,computers and robots.

But where will robots take us, and where will we take them? These are very pertinent questions. With the arrival of radios and then televisions in the twentieth century, extended families ceased to gather around the fireplace to hear ghost stories told by grandpa. Long before that automobiles had replaced the horse driven coach. Long trips that took several days across one's state or months across the entire country became much shorter and frequently more stressful as today we fret over automobile gas costs for a long trip or over canceled airline schedules. Industrialization and mechanization removed the friendly cobbler and the chimney sweep from the scene. Today we have seen globalization turn many local brick and mortar businesses into things quaint and extinct. Smart phones, handheld digital devices and tablets with displays we can manipulate either with a finger or with a wand, chatrooms and social networks? These things in our digital worlds have replaced books, breakfast conversations between family members and reasoned debate. New technologies certainly might not be evils in themselves. Still one must admit that as human technology advances from one generation to the next, human behavior also is transformed and drastically so.

In what ways? To illustrate: A coachman who is intoxicated and driving his horse driven coach in 1867 on a Friday evening in London cannot cause as much damage as an intoxicated accountant can do on one Friday evening in 2016 if he is driving his SUV on some Los Angeles freeway.

Does human behavior improve as human technology advances? Sadly that does not seem to be the case. In every century as technology moves upward toward higher levels of sophistication it seems that humans descend downward to lower levels of moral debasement.

Today some people await the time when robots not only will resemble the best physically attractive looking humans among us, but also will be able to simulate and to act out various human actions, emotions and behaviors. To some extent this already is possible with "embodied conversational agents" or virtual human chatbots, as some people call them. Virtual humans, which can resemble real humans remarkably well sometimes, communicate and interface with real humans through AIML. Perhaps some humans might imagine the fun it would be to romp in bed with some stunningly attractive but totally mechanical sexpot. The wiser humans among us though might pause to consider the possible evil outcomes of such a technological development. We know already what malicious software can do to computer networks, yet this only is some computer code running on an operating system. It would be disastrous for some future human being in some position of great responsibility such as a military officer, a senator, state department or foreign ministry official or corporate executive, to be at a party or public gathering, then lured into a sexual encounter with an attractive, human looking android. An enemy foreign power easily could control the whole scenario for the purpose of blackmail, assassination or worse.

If he was alive today the planetary astronomer and very perspicacious humanitarian Carl Sagan might be compelled to admit that "Spaceship Earth" is headed for a fatal collision with a worldwide ship of fools. Various human factions, ever intractable in their differing political, religious, social and racial ideologies, continue still after time measured in millenia to close ranks, human group population aligned in hostility against some other human group population. After all is it not always "those other people" who are responsible when things go awry? So there is nothing else for humans to resort to when things go wrong either in a country or in the world but to tribalism, nationalism and partisan bickering. No one today seems to have the wisdom, the will, the common sense or even the intelligence to stop the ships from crashing.

In Nigeria, in Syria and in Iraq, Boka Haram and ISIS members insist that Allah the merciful has given them the right to pillage cities and towns, to rape little girls and their mothers, to murder archaeologists and scholars such as Doctor Khaled al Assad of Syria and to destroy precious ancient artifacts that date back to the earlier millennia of Assyrian kings Sennacherib, Esarhaddon and Ashurbanipal and to the later times of General Pompey of the Roman Republic. Yet they claim the West is evil. In the United States politics has become the politics of rancor, divisiveness, showmanship, race bigotry, calumny and slander. Many Christians in name at least, particularly some among those who enforce the law and even some citizens themselves whether of the upper middle class, or poor and members of minorities, clamor ever for justice while simultaneously they pursue injustice with an intense and ardent fervor. In Europe there has been a rise in extremist and xenophobic right wing ideologies such as in France and Austria. Ironically these people claim that Muslims and foreigners from the Middle East and North Africa are evil. *An eye for an eye leaves the whole world blind,* said Mohandas Ghandi.

When today one takes a cold and objective look at the present world of humanity, we do not see anywhere within our restless, planet wide society of high speed wireless communications, SCADA systems, bioengineering and jumbo jets, a superior race of Nietzschean Overmen that has surpassed in their Darwinian evolution the entire human race. Rather we see instead various countries and societies worldwide including here in the West, that are fraught with political, nationalist, religious and ethnocentric pathologies so shameful, destructive and alarming one wonders if *H. sapiens* really will survive the twenty first century without suffering a catastrophic worldwide Internet disaster or an outright thermonuclear war.

Are human beings today so much wiser than were Dwight D. Eisenhower, John F. Kennedy and Nikita Khrushchev in 1962, Mohandas Ghandi, Albert Schweitzer and Martin Luthor King, or are humans today more egotistical and self–righteous? In their century they managed to avoid an H–bomb war and resisted human selfishness and bigotry. In contrast today it looks as if this world's leaders in politics and in business are racing toward the fatal precipice along with all the populists.

Blame whomever one wishes when "things go wrong." The only things on Earth that might survive a sudden and abrupt nuclear war caused by human intractability and self-righteousness will be the Cesium 137 and Strontium 90 isotope residue that will pollute the atmosphere[1], along with perhaps, some sturdier little tardigrades. Should that occur either soon or in the distant future due either to political tensions in Ukraine, the Middle East or in the South China Sea, *what good can artificial intelligence and robots do for us then?*

So have we seen in actuality the ascendancy of Nietzsche's future *Übermenschen* as was suggested in the science fiction movie classic *2001: A Space Odyssey,* or instead humanity's descent into a moral abyss by means of some overly enthusiastic collective human thirst for recidivism and self destruction, back toward the crude morality of *Australopithecus boisei*? These are important questions, for Mary Wollstonecraft Shelley's Frankenstein monster was created in humanity's image. Its human creator was horrified not only by the physical repulsiveness of his creature but also in the sheer delight the creature took in killing. But what did Frankenstein's monster do that humans had not done before its creation and even after its creation? In fact what has been far more physically repulsive over the last century, what Viktor von Frankenstein's monster looked like,

---

[1]Researcher Richard L. Garwin discusses the shaky relationship between peace, science and political power in the essay, "The Relationship of Science and Power." This essay is found in the book *Carl Sagan's Universe.*

or two world wars, Auschwitz-Birkenau, Hiroshima, the murder of Emmet Till, My Lai, Khmer Rouge and September 11, 2001?

In the play *Man and Superman*, George Bernard Shaw suggests that woman is for man some sort of *élan vital*, a driving force that urges him on toward creativity and personal improvement. Well one can ask, if Adolph Hitler was a monster, was Eva Braun a saint? If she was a positive driving force in the life of Hitler, how does one characterize the kind of personal improvement she inspired within him? My point is that men and women *both* have the capacity and often the willingness even, to defend or to ignore morally evil words and actions, although this is not true for intelligent machines at present. Some feminists have claimed that women are better than are men. But are all women better morally or wiser than are all men? Was the imperialist Queen Victoria wiser than Jesus the Nazarene? Was Bishop Cyril in the Third century AD the ethical and moral superior to Hypatia of Alexandria, the talented Greek mathematician he succeeded in getting murdered at the hands of a frenzied Christian mob?

The stubborn proclivity to over generalize about entire groups of people, which certainly is a trait that is deeply human regardless of one's sex, ethnicity, nation or religion, precisely is among the sorts of things that led to *Kristallnacht*, the outbreak of World War Two, the tragic events at Auschwitz-Birkenau and September 11, 2001.

Within fifty years of the play's performance both men and women worldwide had done little to support the Irish playwright's thesis, in a generation that witnessed two destructive world wars, mustard gas, fighter bombers, aircraft carriers and tanks in warfare, the Holocaust and nuclear missiles. Our technological achievements do much more than just to tell the story about the human inventiveness that led to the development of tanks, thermonuclear bombs, ICBMs, mustard gas, biological weapons and robots. They also tell a great deal about the kind of "intelligent" species that created these things, yes, intelligent perhaps, but certainly not a species of superior *Übermenschen* that has moved far beyond good and evil.

Will robots one day prove themselves to be more ethical, more moral than are humans? In a world that truly is superior to our own, ethics and morality would have no significance any more than would good or evil. Neither an elephant nor the robot Atlas needs holy scripture or a body of manmade laws to tell it how to behave. Animals and machines both conform automatically to natural and physical laws without the need for any modification, distortion or perversion of those laws. In contrast to that all organized religion, moral codes and human laws such as on civil and human rights are like safe playpens for babies. The safe playpen does no good if the baby keeps climbing out of it and the playpen is needed no longer once the child truly has become a wise and responsible adult.

Brute animals and machines are by far less destructive and less hurtful than are human beings. Today many societies that call themselves "civilized" and that have an abundance of laws, institutions and organized religions have also an abundance of moral evil, confirmation bias and dysfunction. In the novella *The Mysterious Stranger*, Mark Twain's main character argues that the brute animal is the true superior to mankind because it lacks the "moral sense." There is an inescapable irony in that neither brute animals nor intelligent robots build fighter jets out of anxiety or fear, maliciously spread lies, slander, misinformation and malware across the Internet or feel any motivation based on greed or hatred to commit the wholesale slaughter of millions of dodo birds, buffalo, passenger pigeons, elephants, rhinoceroses and men and women who suffer the ill favor of some self-righteous human majority. Yet "superior" humans have peace treaties, constitutional laws and various versions of some "bill of rights," laws on civil and human rights, the "United" Nations and various religious writings such as the Bible, the Talmud, the Bhagavad Gita and the Quran. However none of these things have succeeded in putting effective restraints on the more dysfunctional behavior patterns of men and women everywhere.

Therefore since religious tenets and human laws frequently are violated or broken by imperfect human beings whose actions frequently are motivated by things such as human prejudices, egotism and by passionate emotions such as hatred, fear, distrust and by "gut instincts" which never are dependable one hundred percent of the time, there is no effective or dependable means or mechanism in place to enforce impartially civil and human rights laws worldwide in a way that minimizes destructive human behavior and aggression. Recall that in the 1951 science fiction classic *The Day the Earth Stood Still* the alien Klaatu told the world's scientists that his planet had created "a race of robots" that were capable of enforcing laws impartially on his planet. The human race however, has no Gort.

There is no such problem for animals and machines. Animals subject themselves to natural law while robots and even complex software systems are subjected to fault tolerance and to autonomous feedback and control.

Is it possible that cyborgs and intelligent robots could behave far better in the future than humans have done over the last six to ten thousand years?

In Isaac Asimov's *The Caves of Steel* and in *The Naked Sun*, Isaac Asimov's robots were not the ones pursuing the resolution of conflict by means of murder, or maintaining and stubbornly so, two dystopic societies that regarded each other with suspicion and apprehension. So perhaps Nietzsche's *Übermensch* is not destined to be above humankind at all, if humankind never can advance to such a superior level of being.

All this does not mean that machine learning and artificial intelligence are bad things. In comparison automobiles and jumbo jet planes are good things in that these machines enable humans to cover vast distances across a country or around the world in shortened times in comparison to travel on foot or in a sailboat. Yet no one would consider it wise even for the cleverest automobile driver or airplane pilot to operate his or her machine while under the influence of drugs or alcohol, or while the human operator is depressed or suicidal.

There might be a way out or two ways perhaps, that is, out from the insanity of the current world's pathological predicaments at least for some, if some small independent group of wise people in Western countries and elsewhere take action at once. In a speech in 1963 at American University President John F. Kennedy said "Our problems are man-made. Therefore they can be solved by man." Unfortunately nearly fifty-four years after he spoke those words humanity is not solving society's problems but creating more and more of them. Cosmologist Stephen Hawking has suggested that humans ought to move out into space. Given the unpredictability of world events perhaps a grass roots campaign toward that goal should be initiated within the next ten years or sooner than that. If the human race is to survive a sudden 'death by suicide' of this world civilization, then the road to future survival for some lies in space, somewhere else in this solar system. A grass roots organization can begin right now by drafting feasibility studies and proposals on how best to build a self–sustaining space colony, either on Mars or on Titan, for example.

Any self–sustaining space colony will need robots to perform tasks that are too dangerous for human civilization builders: Robots for construction, for mining asteroids and for transportation. We live in a world for which catastrophic events are so unpredictable that, if a sudden thermonuclear war does not break out on a Saturday morning, then it still is possible for astronomers to detect sudden gamma ray bursts from a nearby supernova on a Thursday evening. So at least some responsible people need to act now toward the actual construction of an initial small, self–sustaining space colony in low Earth orbit. Another possibility is research and development of other means of spaceflight, such as spaceships with ion drives, or building large ships that use sails that are propelled by sunlight. The supplies needed to build the sail ships could be launched into low earth orbit by conventional rockets then assembled there by teleoperated robots and haptics technology.

There does exist a big obstacle to this approach. Space missions are costly enough when nations and governments pay the expenses needed. How then can an independent private group of humans or a small aerospace corporation initiate a space program that could lead to offworld colonization in a way that is both economic and affordable? Rockets that use liquid fuel have been the standard for decades but this is very expensive for a small space faring group of adventurers to enact without a budget that would have to be at least in the tens of millions of dollars. One needs to find cheaper means for space missions in the solar system. One possibility is to build space tethers to put various construction supplies into low Earth orbits.

One other possible option for human survival might be the development of "better and smarter humans," cyborgs that are part human, part machine but that, unlike human beings, are able to resist destructive, negative tendencies or emotions that could inhibit them from working toward a common purpose such as offworld space colonization. Already in Chapter 10 we learned that bioengineering and biomechatronics have enabled scientists and medical professionals to create human-machine hybrids. Such a human-machine cyborg might be adapted for long term space travel, and a human computer interface or augmented reality system could enable them to inhibit aggressive tendencies, negative emotions or destructive patterns of thought and behavior as they work together on an offworld space colony.

15.1 Adam's Creatures, or no Future?

It has been more than seventy years since two atomic fission bombs were used to end World War Two. In all that time we have not seen yet the "nuclear winter" that astronomer Carl Sagan warned us against with such eloquence back in the nineteen eighties. Yet international terrorism only has increased and substantially so since the massacre of Israeli athletes in Munich in 1972 and the airplane hijackings of the nineteen seventies. For the last twenty years and even today war, famine, various massacres and disease epidemics have ravaged Africa. Europe has seen the rapid influx of millions of Muslim war refugees and also the rise of racism and xenophobic, nationalistic movements. Meanwhile in recent years the United States has been ravaged by wave upon wave of new and more violent crimes, race riots and looting, disrespect for the civil rights and property rights of others, police brutality, political corruption and political factionalism so extreme that lies, calumny and bigotry seem to have transformed into American virtues. Cosmologist and theoretical astrophysicist Stephen Hawking, when asked a question about humanity's failings, remarked "The human failing I would most like to correct is aggression. It may have had survival advantage in caveman days, to get more food, territory or partner with whom to reproduce, but now it threatens to destroy us all."

Webster's 11th Collegiate Dictionary defines the word "civilization" as "Refinement of thought, manners or taste. A situation of urban comfort." But when overly aggressive citizens in the same civilization have stopped being civil to one another, you have civilization no longer, nor democracy. Democracy does not prevail wherever human aggression and tribalism dominate. Indeed here in the United States true democracy is under a direct threat not by nations abroad but by the disruptive behavior of millions of its citizens at home. At present increasing racial hostilities and hate crimes, anti-Muslim fanaticism and xenophobia are dividing the American citizenry at a time when there is increased hostility against America abroad not only by ISIS but also by nation states like China, Iran and North Korea. Any American citizen who thinks this is the time in America to whip up racial tensions and bigotry, to incite overt race discrimination and hate crimes, to foment religious bigotry and strife at home or to resort to rioting, looting, shopping mall brawls, to the murder of police officers, to acts of arson against mosques, Christian churches or synagogues or to engage in other acts of violence in order to get one's way, either must be bone deep evil at heart or dangerously stupid. Senator Barry Goldwater once issued a warning not to the Soviet Union but to the citizens of his own United States, when he reminded Americans in *The Conscience of a Conservative* that the United States of America can collapse from its own citizens neglecting the Bill of Rights and constitutional rule of law at home just as easily as it can fall from the hostile activities of enemies abroad.

It was disunity, internal dissension and strife (mostly between the many local Soviet Republics with different ethnicities) that caused the collapse of the Soviet Union after 1994, not NATO. It was ethnic strife, nationalism and racial tensions between Czechs, Germans, Slavs, Serbs, Magyars and Croatians that caused the Austro-Hungarian Empire to crumble away into oblivion after 1918, not England and France. Today these same things can cause America to collapse, a fact which this nation's enemies abroad in Syria, Teheran, Beijing and Pyongyang either must know very well already or eventually they will find it out because they might watch the collapse firsthand. Unfortunately we are the ones who do not seem to realize the truth of this.

With regard to the future of America Adolph Hitler once said: "I do not see much future for the Americans. It is a decayed country. And they have a serious racial problem over there, and the problem of social inequalities." Yes, Hitler might have been an evil man, but he did have the very successful gutter instincts of a lynch mob leader. Was he right in his prediction?

No longer are men and women the primitive inhabitants of the Pleistocene epoch, hunting like *Homo erectus* then scraping cooked meat off bones with stone tools. Today we live side by side in a world society permeated with guns, violent and extreme religious and political ideologies, computer malware and thermonuclear weapons. Fifty thousand years or more in our past human aggression, tribalism, race prejudice and hatred, strife and religious bigotry are things that might not have caused the extinction of Homo erectus or Homo habillus who both lived prior to the Bronze Age. Yet today in a world civilization that has both thermonuclear weapons and ransom malware capable of encrypting the software used by governments and financial institutions alike, such negative human traits as hatred, bad manners and aggression in word and action could lead to devastating international disasters that grow worse and worse and one after the other, even if they do not lead to outright human extinction.

Any wise human being right now would ask: If the worst should happen and civilization really is obliterated by a sudden thermonuclear war due to some extreme ideologies or to a foolish and sudden misunderstanding between this world's political leaders, to where can a sensible, tolerant, rational and diverse community of people escape? Almost fifty years after the first lunar landing there is not one single space colony that can sustain the survival of even one thousand human beings, to escape the rabid and pervasive *Stupor mundi* of our times down here on Earth.

Truly intelligent machines never will exist at all if their human makers become extinct before they are created. So it is not AI nor robots that threatens the survival of the human species, but rather it is this world's human bigots, bullies and egotists and their apologists who are doing it.

In *Thus spake Zarathustra*, Zarathustra encounters an old holy mystic in the mountainous woods. He converses with the old seer but then he leaves the fellow alone in the woods and astonished in a sense, wondering that the old saint in the woods never had learned the news that "God is dead." Today in the United States regular Christian church attendance is ever active and vibrant. Every day we are told either that Christianity or Islam makes one a better human being. Evangelical television programs can be found both on digital as well as on cable television channels. One need not take much time searching online to find at least tens of thousands of search links to religion related websites, religious forums, online religious services and sermons, etc. Moreover one sees everywhere throughout America and the world that both Judaism and Islam have millions of very active religious participants.

Every year billions of Christian, Jewish and Muslim men and women happily celebrate Easter, Christmas, Hanukah and Ramadan. Yet despite all this active enthusiasm, vibrancy and devotion to religious faith and practice, the country and the world quake and convulse with nationalism, ethnocentrism, xenophobia, violence, political enmity, political and religious extremism, religious strife, greed, bad manners, arrogance, racial hatred and division.

But just what good is one's race, religion, politics or nation doing to benefit the entire human species? Why is it that religion, democracy, politics and civil and human rights laws have failed to make the world a far safer place than it was back in 1914 or 1939? These things cannot work because human beings driven by unrestrained emotions and by groupthink insist upon breaking moral, legal and ethical rules in the name of one's religion, race, nation, tribe or political party. Politics, laws and religion only are pretty storefronts, to make human beings feel comfortable about the societies in which they live despite what civil rights abuses or perversions of justice might continue to occur within those societies. There is a word to describe all that and the word is hypocrisy. We have become a worldwide human civilization of religious and political hypocrites, bigots, bullies, egotists, liars, pleasure-obsessed power abusers and control freaks. If superior alien civilizations have managed to decipher our radio and video signals somewhere out there within the remote vastness of interstellar space, they must look upon the entire human species with alarm and astonishment. Afterward they might look upon their own past with some relief and satisfaction that they themselves did not allow their own inner evil urges and inclinations to determine the future fate of their civilization.

So one can ask *what really is dead today*, God, or Judeo-Christian ethics and the graciousness and mercifulness of Allah? In truth neither organized religion nor dogmatism of any kind seems to be working to make the world a safer place. Tension and friction repeatedly have assailed human relationships around the world, threatening to choke the life out of civilizations like a heavy weight around the neck, time and time again. That has been true for more than two thousand years since the times and early teachings of

Moses, Buddha, Lao Tze and Jesus the Nazarene.

As we have seen, today some scientists and technologists fret about the future potential dangers of AI. But one asks again and with all due respect to cosmologist Stephen Hawking and to space pioneer Elon Musk, who have expressed alarm about the advancement of AI, what really is the *greater* danger to the future security, safety and perpetuation of Carl Sagan's "Spaceship Earth" replete with its sophisticated biosphere, its variegated life forms and the human species, intelligent robots, or we humans ourselves?

Who in fact is much more dangerous right now to the safety of civilization and to the Earth, men and women or the intelligent machines they use? We have seen human beings develop everything from bronze swords to tanks to ICBMs, thousands of thermonuclear bombs and online malware over the span of thousands of years. To expect that human beings within the next ten or twenty years will develop more advanced artificial intelligence systems and robots that never will be used by humans in destructive or violent actions is wishful thinking.

On the other hand could intelligent machines one day preserve, protect, defend and maintain this "Spaceship Earth," with all its vast, intricate, diverse and interconnected parts, in ways that are much better than we dogmatic humans have done? One wonders. Perhaps they can do so. When it comes to nationalism, xenophobia, ethnic strife or political and religious factionalism, machines are neutral in the purest sense.

After all unlike human men and women, intelligent machines are the perfect agents for logic based decision making and for bias freedom, precisely because they are not human. Unlike human beings they are not born with dark skin or blue eyes. They have no biological genes, no alleles, no evolutionary traits in favor of kin selection or inclusive fitness. Like a *tabula rasa* or like the bare metal hardware on which one builds a brand new operating system for experimentation, every brand new robot or artificial intelligence system is free from all corruption moral or otherwise. So unlike human beings these things do not organize themselves into mutually exclusive, disparate groups fueled with passions of hostility and distrust toward each other based on left wing or right wing politics, culture, nation, tribe, race or religion.

Could it be perhaps that at least some scientists fear that intelligent machines one day will be superior to men and women and to make better decisions than have we humans over the last fifty centuries or so? In essence might they be better than humans one day? Perhaps. Intelligent robots do not vote motivated from decisions based on the political candidate, ethnicity, race or party they hate. Nor do they cyber bully, commit murder, rape or "hate crimes," exhibit bad taste, speech or manners in public, go to Mass, Easter or Passover services only whenever everyone else inside their cathedral, church or temple looks like them and lives in the same neighborhood, or make pilgrimages to Mecca where frequently some devotees might get trampled to death in the midst of a sudden stampede of irrational and emotion-driven worshipers.

The historian Arnold Toynbee once claimed, "The weak spot of religion is its ridiculousness."

In some Bible translations it is said that God is a jealous God. Humans too, made in the image and likeness of God according to many Jews and nominal Christians alike, can be motivated by jealousy. One only can hope that one day, *Adam's creatures* can be created much better than that.

Doubtless some will believe this discussion on human intolerance and moral irresponsibility with the use of new technologies like robots and AI is unnecessarily bleak. They will argue that human conflicts and disagreements can be resolved through debate, through dialogue, by understanding the value in having racial, ethnic, religious, cultural or national differences, to see that actually some commonality lies underneath them, etc. But human dialogue and debate between in-group and out-group populations never prevented the American Civil War. Such actions never prevented the roar of guns and the spread of carnage and pandemic after August, 1914. Nor did they stop the world from going to war once again after the Munich Agreement in 1938. Dialogue and debate via telephone between Attorney General Robert Kennedy in Washington DC and Governor George Wallace in Alabama never brought an end to racial hatred and race-based murders. It was not dialogue between Secretary of State Henry Kissinger and the Viet Cong in Paris, France that brought a consummation to the war in Vietnam, but rather the fall of Saigon. Moreover dialogue and discussion between Prime Minister Mechaem Begin from Israel, President Anwar Sadat from Egypt and President Jimmy Carter of the US, never ended terrorism and conflict in Palestine. To such optimistic individuals who still believe that the human species can overcome various differences that have caused violence, distrust and hatred between humans for thousands of years, Oswald Spengler (who had warned Europe that a more devastating war was imminent after 1918), also would have warned: "Only dreamers believe in a way out. Optimism is cowardice."

In ten thousand years humankind has not reached yet even the level of a Type I civilization, based on the Kardeshev scale, despite the discoveries of nuclear fission, neutron stars, machine learning and Moore's Law. So perhaps there is no hope that contemporary humanity and its posterity as a whole will advance into the future to become a Type I civilization. But that does not mean each and every individual in the future will not find some way to escape the various forms of collective madness that has plagued humanity continually for millennia. If the worst should happen so that a greater portion of humanity does not avoid annihilation in yet a Third World War, perhaps some small remnant of humans then will have learned the hard lessons at last, to leave the confines of Earth for elsewhere, along with robots.

How quickly could a Type I civilization of highly intelligent machines and cyborgs progress to the level of a Type II then to a Type III civilization? One wonders. Perhaps over the next ten thousand years they could impose some *Pax Robotica*, to supervise and to maintain planet Earth along with its future wildlife and biosphere (just like Dewey, the little service robot, cared for the last remaining fauna and flora from Earth toward the end of the film *Silent Running*), doing all this far better than emotion-driven and continually feuding humans have done over the last millenia.

Appendix

15.2 Some Areas in Mathematics related to Robots and Machine Learning

This Appendix is for any readers who, although not having a background in robotics, in computer science or in mathematics might have had some calculus or physics in high school or in college. To provide a comprehensive tutorial or primer on robotic kinematics, such as one might need to build computer simulations and animations of various robot actuators with a physics engine is beyond the scope of this book. For that one will have to consult mechanical engineering texts on robotics or do a search online with the keywords "Robotic kinematics," or "inverse kinematics." However in this Appendix the reader will find some background material on classical mechanics and discrete probability that serve as information one needs to know even before one begins to study robotic kinematics, artificial intelligence or machine learning.

15.3 Kinematics and Dynamics

Let $\mathbf{x}$ be a vector on $\mathbb{R}^2$ or on $\mathbb{R}^3$. By Newton's Laws of Motion the velocity of a point mass particle is denoted in Cartesian coordinates by

$$\mathbf{v} = \dot{\mathbf{x}} = \frac{d\mathbf{x}}{dt}. \tag{15.1}$$

Its acceleration is given by

$$\dot{\mathbf{v}} = \ddot{\mathbf{x}} = \frac{d^2\mathbf{x}}{dt^2}. \tag{15.2}$$

From this we get the related momentum and Force related to the particle with mass m respectively as

$$\mathbf{p} = m\dot{\mathbf{x}} = m\frac{d\mathbf{x}}{dt} \tag{15.3}$$

and

$$\dot{\mathbf{p}} = m\dot{\mathbf{v}} = m\frac{d^2\mathbf{x}}{dt^2}. \tag{15.4}$$

When $\frac{d^2\mathbf{x}}{dt^2} = \mathbf{g}$ with $\mathbf{g}$ being acceleration due to gravity the particle has weight $m\mathbf{g}$. From all this we can get a second order differential equation

$$\frac{d^2\mathbf{x}}{dt^2} - \frac{F}{m} = 0. \tag{15.5}$$

Solutions to this differential equation gives the particle's trajectory either in two or three dimensions.

Sometimes it is necessary to express things in other coordinate systems, such as in polar coordinates. When that is the case we get $x = r\cos\theta, y = r\sin\theta$ and equations for velocity and acceleration as

$$\mathbf{v_{r,\theta}} = \dot{\mathbf{r}} = \dot{r}\mathbf{e}_r + \dot{\theta}r\mathbf{e}_\theta. \tag{15.6}$$

The vector components of $\ddot{\mathbf{r}}$ become

$$\ddot{r} - r\dot{\theta}^2, \ddot{\theta}r + 2\dot{r}\dot{\theta}, \tag{15.7}$$

where $\mathbf{e}_r$ and $\mathbf{e}_\theta$ are the unit vectors in polar coordinates and

$$\frac{d\mathbf{e}_r}{dt} = \mathbf{e}_\theta\frac{d\theta}{dt}, \frac{d\mathbf{e}_\theta}{dt} = -\mathbf{e}_r\frac{d\theta}{dt}. \tag{15.8}$$

The work done by a force vector $\mathbf{F}$ on a particle moving along some curve is an integral dot product

$$W = \int_{r_1}^{r_2} \mathbf{F} \cdot d\mathbf{r}. \tag{15.9}$$

Since

$$d\mathbf{r} = \frac{d\mathbf{r}}{dt}dt, \tag{15.10}$$

one can show the work actually is the kinetic energy between two times t_1 and t_2 where $t_2 \geq t_1$,

$$W = \frac{1}{2}m\left(v_{t=t_2}^2 - v_{t=t_1}^2\right). \tag{15.11}$$

There is a function of position and velocity called the "Lagrangian" $L(q, \dot{q}, t)$ where q is a coordinate and $\dot{q}$ is velocity, that enables one easily to obtain the equations of motion of a particle in a wide variety of different kinds of problems. When the force is conservative the relationship between L and p is

$$p = \frac{\partial L(q, \dot{q}, t)}{\partial \dot{q}}, \tag{15.12}$$

such that

$$\frac{d}{dt}\frac{\partial L(q, \dot{q}, t)}{\partial \dot{q}} - \frac{\partial L(q, \dot{q}, t)}{\partial q} = 0, \tag{15.13}$$

where $L = T - U$ and T and U are kinetic and potential energy respectively. Solving this "Lagrangian equation" derives the equations of motion for the particle. There also is another important function one might want to learn about called the Hamiltonian, which we do not consider here.

15.4 Rigid Body Motion

A rigid body is a mass object such that the distance between any two particular points in it or on it are constant when the whole object moves by translation or by rotation. Some examples of rigid bodies would be wooden blocks, bowling balls, steel cylinders, tables, chairs and the different parts of a robotic arm for instance, that is, each part that does not move although there are movements at the joints. But one also can include the platform on which the robotic arm is attached as being a rigid body. Some examples of things that are not rigid bodies would include water inside a cup or the air inside a room.

Suppose a solid steel drum or solid cylinder is rotating about an axis of symmetry that passes through its center of mass. The drum has an angular momentum

$$\mathbf{\Omega} = m(\mathbf{r} \times (\omega \times \mathbf{r})) = \mathbf{I}\omega, \tag{15.14}$$

where ω is the angular velocity about the axis through a fixed point and $\mathbf{I}$ is the moment of inertia.

The cylinder also has a rotational kinetic energy

$$\frac{1}{2}\omega \cdot \mathbf{\Omega}. \tag{15.15}$$

Using $\mathbf{\Omega} = \mathbf{I}\omega$ one can express this as

$$\frac{1}{2}\omega \cdot \mathbf{\Omega} = \frac{1}{2}I\omega^2. \tag{15.16}$$

The cylinder's motion has six degrees of freedom found at the center of mass, three for translation and three more for rotation. For example the cylinder could be a robot probe moving through space and rotating around an axis of symmetry as it does so. Frequently there are three angles called "Euler angles" that help mechanical engineers perform calculations involved with rigid body motion within two different coordinate systems, where one of these coordinate systems is fixed on the rigid body and the other one is fixed in space.

15.5 Quaternions

Computations in robotics frequently involve coordinate transformations, linear transformations from one coordinate system to another coordinate system. These can be either for translations or rotation. Also one needs to use certain kinds of projective spaces where one uses "homogeneous" coordinates. In some settings quaternions are useful for this, especially when one combines work in robotics with computer graphics.

A quaternion is a four dimensional complex number

$$a + bi + cj + dk, \tag{15.17}$$

$$a, b, c, d \in \mathbb{R}, \tag{15.18}$$

where

$$i^2 = j^2 = k^2 = l^2 = \sqrt{-1}. \tag{15.19}$$

There is a multiplication rule

$$ij = k, jk = i, ki = j, \tag{15.20}$$
$$ji = -k, kj = -i, ki = -j. \tag{15.21}$$

Just as the integers are a subset of the rational numbers and the rational numbers in turn are a subset of the reals, the reals are a subset of the complex numbers and the complex numbers are a subset of the quaternions. In turn even the quaternions themselves are a subset of a larger set of complex numbers called the octonions in an eight dimensional complex space.

One can add together two or more quaternions q_1, q_2, to get another quaternion

$$q_3 = q_1 + q_2. \tag{15.22}$$

One also can multiply two or more quaternions together to get another quaternion. However unlike the multiplication of complex numbers, multiplication of quaternions is not commutative. So one will find in general that

$$q_1 q_2 \neq q_2 q_1 \tag{15.23}$$

holds in general, if q_2 is not the complex conjugate of q_1. Each quaternion not equal to zero has an inverse,

$$q^{-1} = \frac{q^*}{|q|^2}, \tag{15.24}$$

where q^* is the complex conjugate of q. Suppose $|qq^*| = 1$. Then $q^{-1} = q^*$ which is useful since the inverse for a rotation matrix A is such that $A^{-1} = A^T$, where A^T is the transpose of A. Barely have we scratched the surface of the topics one needs to know to be familiar with what role quaternions and rigid body motions play in robotics, in part because one needs to be more familiar with linear algebra and matrices than just calculus, topics outside the scope of this book. But one can read much more extensive material about quaternions, rigid body motions and rotation matrices in *Quaternions and Rotation Sequences*, by Jack B. Kuipers. Also there is a very good and extensive tutorial "Representing Attitude: Euler Angles, Unit Quaternions and Rotation Vectors" by James Diebel (See the Bibliography) at

http://www.swarthmore.edu/NatSci/mzucker1/e27/diebel2006attitude.pdf.

15.6   Homogeneous Coordinates

Frequently mechanical engineers use homogeneous coordinates when they make calculations related to various linear transformations that have to do with coordinate translations and rotations. One coordinate system that helps is called "homogeneous" and has to do with projective spaces, something that also is used frequently by people who work on virtual reality programs and in the computer graphics industry.

In the two dimensional space of complex numbers (mathematicians usually denote it by $\mathbb{C}$) one can find a 1-1 rule of assignment from each point on the unit sphere onto the complex plane, where every point gets mapped to some point on the plane and the "North Pole" gets mapped to what one calls the "point at infinity." The sphere for the mapping is called the Riemann sphere and the map is known as a "stereographic projection."

But one can do something similar with real spaces like $\mathbb{R}^3$ and $\mathbb{R}^2$. For instance the point "at infinity" on $\mathbb{R}^2$ gets mapped to the point $(0, 0, 1)$ on $\mathbb{R}^3$. This helps to enable one to identify different points as being identical when they are in some projective space. Let λ be any nonzero real number and

$$(x, y, z),$$

some point in $\mathbb{R}^3$ not the origin where x, y and z are not all equal. Then ordinarily the points (x, y, z) and $(\lambda x, \lambda y, \lambda z)$ will not be the same. But in the projective space $\mathbb{RP}^3$ we get an equivalence

$$(x, y, z) \sim (\lambda x, \lambda y, \lambda z),$$

which means the two points (x, y, z) and $(\lambda x, \lambda y, \lambda z)$ are the same in the projective space $\mathbb{RP}^3$ when we consider all the points in

$$\mathbb{R}^3 - \{(0, 0, 0)\},$$

that is, all points except for the origin.

Homogeneous coordinates are useful in robotics and computer graphics because for one thing it helps to reduce the dimensionality of some linear transformations from three dimensions to two dimensions.

A projective space has some unusual properties. For instance in a projective space it can be true that given a line and a point not on the line, two or more lines can pass through that point while being parallel to the given line[2]. Also under certain conditions a line can "become" a circle and a parabola an ellipse, when one includes the "point at infinity."

15.7 A Short Primer on Discrete Probability

Let k be a large positive integer. Suppose someone performs an experiment that has a total number of k sample items possible in a "sample space" S, such that the experiment can have different kinds of outcomes and events which are subsets of the sample space. For example if the sample space is

$$S = \{Heads, Tails\}, \tag{15.25}$$

we have $k = 2$ and the possible events

$$\{Heads\}, \{Tails\}. \tag{15.26}$$

If the experiment has to do with what face shows upward after tossing a fair die, then

$$S = \{1, 2, 3, 4, 5, 6\} \tag{15.27}$$

and $k = 6$. There are many other experiments, sample spaces and events possible. Not all of them are quantifiable so easily, for instance how many people share the same birthday in a room of one hundred people.

There is a rule of assignment, call it "P" between the event subsets in a sample space and subsets on the real line. So if a particular event is a subset

$$A \subset S, \tag{15.28}$$

where $A \neq \emptyset$, the result can be

$$0 \leq P(A) \leq P(S), \qquad . \tag{15.29}$$

---

[2]This peculiar property is a characteristic of the geometry Albert Einstein used in his Special Theory of Relativity, where parallel lines (geodesics) are circular arcs on the Poincaré disk, which has a hyperbolic geometry.

since $|A|$ which is the cardinality of the set A, can be no larger than the cardinality $|S|$ of S. Clearly we have a way of quantifying an event since every subset of S must have cardinality equal to or smaller than $|S|$:

$$P(A) = \frac{|A|}{|S|}. \qquad (15.30)$$

Clearly we have also

$$P(S) = \frac{|S|}{|S|} = 1, \qquad (15.31)$$

$$P(\emptyset) = \frac{|\emptyset|}{|S|} = 0. \qquad (15.32)$$

So we understand $P(A)$ to be 'the probability of the event A."

Now we introduce another concept called a "random variable." A sample space does not have to have numbers in it. For instance the sample space in Eqn. (15.26) has elements "Heads" and "Tails," which are not numbers. Also for the toss of a die the elements in S are six numbers but really they are labels on the faces of a single die. To deal with this we call X a random variable if for every a in some subset A in S we can crank out a real number x on the real line, such that

$$X(a) = x. \qquad (15.33)$$

Now we see that if $A = \{Heads\}$ say, then $a = Heads$ and $X(a) = X(Heads)$ is some real number. Now we can get probabilities by associating subsets and elements from the sample space S with subsets and elements on the real line $\mathbb{R}$, since

$$P_X(A) = P(a : X(a) \in \mathbb{R}, a \in A), \qquad (15.34)$$

depends upon

$$X(a) \in \mathbb{R} \qquad (15.35)$$

even when $a \notin \mathbb{R}$. Thus if S is the sample space $\{Heads, Tails\}$ and $A = \{Heads\}$,

$$P(A) = \frac{|A|}{|S|} = \frac{1}{2}, \qquad (15.36)$$

so that for example we can take when $a \in \{Heads\}$,

$$X(a) = X(Heads) \in \{0, 1\} \subset \mathbb{R}, \qquad (15.37)$$

where this means $X(a) = 1$ if $a \in \{Heads\}$ and $X(a) = 0$ if $a \in \{Tails\}, a \notin \{Heads\}$.

Now suppose one has a hat that contains 55 green marbles and 45 red marbles. Then the sample space is not numbers but one hundred green and red marbles. Let A be the subset of green marbles and B the subset of red marbles with $a \in A, b \in B$. Then with $X(a) = 0$ and $X(b) = 1$,

$$P(A) = \frac{55}{100} \qquad (15.38)$$

and

$$P(B) = \frac{45}{100}. \qquad (15.39)$$

This way we get real numbers $P(A)$ somewhere between zero and one and some other real numbers $X(a)$ for the random variable, even when neither the element a, the subset A nor its superset S are numbers or sets of numbers. So this is useful if we want to know if a coin shows Heads or Tails, if the number that shows up on the face of a fair die is even or odd. But this characterization lurks in the background when a data scientist is trying to study massive data sets to determine whether someone with heart disease is older than forty or younger than forty or if a given handwritten character is a "C" or an "O."

Here are some axioms from discrete probability that need to be satisfied for any subset $A \subset S$:

1. $P(\emptyset) = 0$,

2. $0 \leq P(A) \leq P(S) \ \forall \, A \subset S$,

3. $P(S) = 1$.

We also have some other results, for example

$$
\begin{aligned}
P(A) &= 1 - P(A^C), & (15.40) \\
P(A^C) &= 1 - P(A), & (15.41)
\end{aligned}
$$

where A^C denotes the set in a sample space S that is the complement of the set A. So for instance if the sample space is $\{Heads, Tails\}$ and A is $\{Heads\}$, then

$$
A^C = \{Tails\} \tag{15.42}
$$

While for the faces of the die we would have if $A = \{2, 4, 6\}$,

$$
A^C = \{1, 3, 5\}. \tag{15.43}
$$

Now let S be some other discrete sample space and A, B two different subsets. We can define a "conditional probability"

$$
P(B|A) = \left(\frac{P(B \cap A)}{P(A)} \right). \tag{15.44}
$$

This is the probability that event B occurs given that event A has occurred. However whenever the events A and B are *independent* we get that

$$
\begin{aligned}
P(B|A) &= \left(\frac{P(B \cap A)}{P(A)} \right) & (15.45) \\
&= \left(\frac{P(A)P(B)}{P(A)} \right) & \\
&= P(B). & (15.46)
\end{aligned}
$$

Let A be some event from some sample space S such that $P(A) > 0$ and made up of n mutually exclusive subsets

$$
A = \cup_{i=1}^{n} A_i, \tag{15.47}
$$

where by "mutually exclusive" as subsets we mean

$$A_i \cap A_j = \emptyset \ \forall i, j \in [1, n], \ i \neq j. \tag{15.48}$$

Then

$$A = \cup_{i=1}^n (A \cap A_i) \Longrightarrow P(A) = \sum_{i=1}^n P(A|A_i)P(A_i), \tag{15.49}$$

by which we can get

$$P(A_i|A) = \frac{P(A_i \cap A)}{P(A)} = \frac{P(A|A_i)P(A_i)}{\sum_{i=1}^n P(A|A_i)P(A_i)}, \tag{15.50}$$

which we call *Bayes's theorem*.

Conditional probabilities and Bayes's theorem turn up over and over again inside various machine learning algorithms that have the right training data, for example in Bayesian networks and for Naive Bayesian classifiers.

There are some other quantities that are important too in discrete probability such as the mean μ, the variance σ^2, the standard deviation σ and the correlation

$$\rho \in [-1, 1] \subset \mathbb{R},$$

when one is trying to understand some underlying significance inherent within a data set or when one is interested in what kind of probability distribution the set has. One might be surprised to know that some more conventional statistical models are related to machine learning and artificial intelligence algorithms. For instance in some situations linear regression models in statistics can be reformulated as artificial neural networks or perceptrons.

Bibliography

The ALICE and Captain Kirk chatbots and AIML interpreters, available at http://www.alicebot.org/aiml.html.

G. W. Allport, *The Nature of Prejudice*, Doubleday and Company, NY, 1958.

AlphaGo: Mastering the ancient Game of Go with Machine Learning, *Google Research Blog*, January, 2016. Available at http://research.googleblog.com/2016/01/.

J. Arquilla, The Dangers of Military Robots, the Risks of Online Voting, *ACM Communications*, **58** 7 (2015) pp. 12–13.

I. Asimov, *The Robots of Dawn*, Doubleday and CO., Inc., NY, 1983.

—, *Robots and Empire*, Doubleday and CO., Inc., NY, 1985.

T. Chan, A statistical Analysis of the Towers of Hanoi Problem, *Internat. J. Comput. Math.*, **28**, pp. 57–65.

Charles Babbage: Ada Lovelace, available at http://www.computerhistory.org/babbage/adaLovelace.

A. Kar, Lady Ada Lovelace and the Analytical Engine, available at http://people.maths.ox.ac.uk/kar/AdaLovelace.html.

M. deBerg, O. Cheong, M. van Kreveld, M. Overmars, *Computational Geometry: Algorithms and Applications*, Third Edition, Springer-Verlag, Berlin, 2008.

P. G. Bergman, *Introduction to the Theory of Relativity*, Dover Publications, NY, 1976.

C. M. Bishop, *Pattern Recognition and Machine Learning*, Springer, NY, 2006.

Blade Runner (Science fiction film, 1982), Wiki article available at http://en.wikipedia.org/wiki/Blade_Runner.

Brain Computer Interface, an article available at http://helab.umn.edu //eegsensing.htm.

A Brief Academic Biography of Marvin Minsky, available at http:// web.media.mit.edu/m̃insky/minskybiog.html.

A. G. Bromley, Charles Babbage's Analytical Engine 1838, *IEEE Annals of the History of Computing*, **4** (3) pp. 197–217.

M. Burrow, *Representation Theory of Finite Groups*, Dover Publications, NY, 1993.

M. Campbell, A. J. Hoane, Feng Hsiung Hsu, Deep Blue, available at http://sjeng.org/ftp/deepblue.pdf.

M. Caidin, *Cyborg*, reader commentaries and review at http://www.goodreads.com/book/show/1368516.Cyborg. Martin Caidin, aviator, science fiction author. Wiki article at http://en.wikipedia.org/wiki/Martin_Caidin.

The Chinese Room Argument, available at http://plato.stanford.edu/entries/chinese-room/.

Edgar F. Codd, Wiki article available at http://en.wikipedia.org/wiki/Edgar_F._Codd

P. J. Cohen, *Set Theory and the Continuum Hypothesis*, Dover Publications, NY, 2008.

Computational Complexity of Games and Puzzles, http://www.ics.uci.edu/ẽppstein/

cgt/hard.html.

A. Datta, M. C. Schantz, A. Datta, Automated Experiments on Ad Privacy Settings, *Proceedings on Privacy Enhancing Technologies*, **1** (2015).

C. Davies, BApasswd: A new proactive Password Checker, *Proceedings of the 16th National Computer Security Conference*, September 1993.

M. Davis, *Computability and Unsolvability*, Dover Publications, NY, 1982.

J. W. Dawson, Gödel and the Limits of Logic, *Scientific American*, **280**, 6 (1999), pp. 66–81.

The Day the Earth stood Still (Science fiction film, 1951), Wiki article available at http://en.wikipedia.org/wiki/The_Day_the_Earth_Stood_Still.

P. K. Dick, *Do Androids Dream of Electric Sheep?*, Doubleday, NY, 1968.

J. Diebel, Representing Attitude: Euler Angles, Unit Quaternions and Rotation Vectors, available at http://www.swarthmore.edu/NatSci/mzucker1/e27/diebel2006attitude.pdf.

T. G. Dieterrich, E. J. Horvitz, Rise of Concerns about AI: Reflections and Directions, *ACM Communications*, **58** 10 (2015) pp. 38–40.

J. M. Dillard, *Star Trek: Where no one has gone before, a History in Pictures*, Pocket Books, NY, 1994.

Dulcimer, entry available at http://encyclopedia2.thefreedictionary.com/dulcimers.

C. Edwards, Self–Repair Technologies Point to Robots that design Themselves, *ACM Communications*, **59**, 2 (2016) pp. 15–17.

J. Emspak, How a Machine Learns Prejudice: Artificial intelligence picks up Bias from Human Creators, not from cold, hard Logic, in *Scientific American* online, December 29, 2016, available at http://www.scientificamerican.com/article/how-a-machine-learns-prejudice/.

M. Ford, *Rise of the Robots: Technology and the Threat of a Jobless Future*. More information about the book is available at http://econfuture.wordpress.com/about/.

A. S. Fraenkel, M. R. Garey, D. S. Johnson, T. Schaefer, Y. Yesha, The complexity of checkers on an N*N board, preliminary report, *Proc. 19th IEEE Symp. Found. Comp. Sci.*, (1978) pp. 55-64.

N. J. Freundlich, Brain Style Computers, *Scientific American*, February, 1989, p. 69.

D. Geer, Resolved: The Internet is no Place for Critical Infrastructure, *ACM Communications*, **56** 6 (2013) pp. 48–53.

E. Gibbon, *The Decline and Fall of the Roman Empire*, Harcourt, Brace and Co., NY, 1960.

E. Gibney, Google AI Algorithm masters ancient Game of Go (2016), available at http://www.nature.com/news/.

B. M. Goldwater, *The Conscience of a Conservative*, BN Publishing (www.bnpublishing.com), 2007.

G. W. F. von Hegel, *Vorlesungen über die Philosophie der Weltgeschichte*, available at http://socserv.mcmaster.ca/econ/ugcm/3113/hegel/history.pdf.

Hugh Herr, MIT professor in Biomechatronics, his website available at http://www.media.mit.edu/people/hherr.

A. Hinz, The Tower of Hanoi, *L'Enseignement mathématique*, **35** (1989), pp. 289–321.

J. J. Hopfield, Neural Networks and Physical Systems with Emergent Collective Computational Abilities, *Proc. Nat. Acad. Sci.*, USA 79 (1982), pp. 2554–58.

J. P. Hogan, *Mind Matters: Exploring the World of Artificial Intelligence*, Ballantine Publishing, NY, 1997.

A. Hyman, *Charles Babbage: Pioneer of the Computer*, Oxford University Press, 1982.

IBM 7094 (Project MAC), information available at http://mit.edu/6.933/ www/Fall2001/AILab.pdf.

IBM 7094, article entry available at http://www.computerhistory.org/collections/catalog/X837.87.

The IBM 7094, available at http://www.columbia.edu/cu/computinghistory/7094.html.

K. Johns, T. Taylor, *Professional Microsoft Robotics Developer Studio*, Wiley Publishing, Inc., IN, 2008.

G. Joos, *Theoretical Physics*, Third Edition, Dover Publications, NY, 1986.

M. Kantarzdic, *Data Mining: Concepts, Models, Methods and Algorithms*, Second Edition, John Wiley and Sons, Inc., 2011.

K. Kirkpatrick, The Moral Challenges of Driverless Cars, *ACM Communications*, **58** 8 (2015) pp. 19–20.

Kirkwood Gap, article available at http://en.wikipedia.org/wiki/Kirkwood_gap.

Kirkwood Gaps, article available at http://scienceworld.wolfram.com/astronomy/ KirkwoodGaps.html.

D. Kirkwood, *The Asteroids, or the Minor Planets between Mars and Jupiter*, 1888, available at http://www.gutenberg.org/files/41570/41570-0.txt

S. C. Kleene, Lambda Definability and Recursiveness, *Duke Mathematical Journal*, **2** 2 (1936) pp. 340-353.

J. B. Kuipers, *Quaternions and Rotation Sequences: A Primer with Applications to Orbits, Aerospace and Virtual Reality*, Princeton University Press, NJ, 1993.

R. Lidl, H. Niederreiter,*Introduction to Finite Fields and their Applications*, Revised Edition, Cambridge University Press, NY, 1994.

S. Lipschutz, M. L. Lipson, *Discrete Mathematics*, Second Edition, Schaum Outline Series, McGraw Hill, NY, 1997.

S. Lloyd, *Programming the Universe: A Quantum Computer Scientist takes on the Cosmos*, Vintage Books/Random House, NY, 2006.

—, Ultimate physical Limits to Computation, ArXiv preprint, available at http://arxiv.org/ abs/quant-ph/9908043.

T. Logsdon, *Orbital Mechanics: Theory and Applications*, John Wiley and Sons, Inc. NY, 1998.

Ada Lovelace, Wiki article available at http://en.wikipedia.org/wiki/Ada_Lovelace.

C. D. Manning, H. Schütze, *Foundations of Statistical Natural Language Processing*, MIT, MA, 1999.

J. McCarthy, Recursive Functions of symbolic Expressions and their Computation by Machine, http://jmc.stanford.edu/articles/recursive.html.

—, E. Feigenbaum, Arthur Samuel: Pioneer in Machine Learning, *AI Magazine*, **11** 3 (1990). available at http://www.aaai.org/ojs/index.php/aimagazine/article/view/840/758.

M J. McCavitt, Guide to the Papers of Norbert Wiener, Massachusetts of Technology Library, available at http://libraries.mit.edu/archives/research.collections/collections-mc/pdf/mc22.pdf.

S. McClure, J. Scambray, G. Kurtz, *Hacking Exposed: Network Security Secrets and Solutions*, McGraw Hill, NY, 2012.

G. Meyrink, *Der Golem* (1914 novel, German language), available at http://www.zeno.org/Literatur/.

G. C. McVittie, *General Relativity and Cosmology*, University of Illinois Press, Urbana,

Ill., 1965.

S. B. Michaels, List of Oral Histories, available at http://www.crmvet.org/nars/orallist.htm.

M. L. Minsky, *Computation: Finite and Infinite Machines*, Prentice Hall, Englewood Cliffs, NJ, 1967.

M. L. Minsky, A Framework for representing Knowledge, MIT, AI Laboratory Memo 306, June, 1974. http://web.media.mit.edu/m̃insky/papers/Frames/frames.html.

M. L. Minsky, S. Papert, *Perceptrons*, MIT Press, MA, 1969.

M. L. Minsky, *The Society of Mind*, Simon and Schuster, NY, 1986.

—, Will Robots inherit the Earth? *Scientific American*, October, 1994, p. 87.

H. Moravec, *Robot: Mere Machine to Transcendent Mind*, Oxford University Press, NY, 1999.

—, The Stanford CART and the CMU Rover, *IEEE Proceedings*, **71**, 7 (1983).

A. Newell, H. A. Simon, Human Problem Solving, available at http://books.google.com/books/about/Human_problem_solving.html.

—, J. C. Shaw, H. A. Simon, The Logic Theory Machine, *IRE Transactions on Information Theory* (1956).

—, Physical Symbol Systems, *Cognitive Science*, **4**, 2 (1980) pp. 135-183.

NASA's space missions. Information on these missions can be found at http://www.jpl.nasa.gov/missions/.

M. Okuda, D. Okuda, *Star Trek Chronology: The History of the Future*, Pocket Books, NY, 1993.

Publius Ovidius Naso, *Metamorphoses*, available at http://ovid.lib.virginia.edu/trans/Ovhome.htm.

S. Papert, M. L. Minsky, *Perceptrons: An Introduction to Computational Geometry*, MIT Press, MA, 1988.

J. M. Pasachoff, *Astronomy: From the Earth to the Universe*, Fourth Edition 1993 Version, Harcourt Brace College Publishers, NY, 1993.

R. Penrose, *The Emperor's New Mind*, Oxford University Press, NY, 1989.

P. J. E. Peebles, *Principles of Physical Cosmology*, Princeton University Press, NJ, 1993.

J. R. Pierce, *An Introduction to Information Theory: Symbols, Signals and Noise*, Second Revised Edition, Dover Publications, NY, 1980.

—, Language and Machines: Computers in Translation and Linguistics, Publication 1416, National Academy of Sciences, National Research Council, Washington, DC, 1966.

A. Rand, *The Virtue of Selfishness*, Signet, NY, 1964.

Appollonius Rhodius, *The Argonautica*, epic poem translated by R. C. Seaton (1912), available at www.gutenberg.org.

K. A. Robbins, S. Robbins, *Unix Systems Programming: Communication, Concurrency, and Threads*, Prentice Hall, NJ, 2003.

C. Robert, G. Casella, *Monte Carlo Statistical Methods*, Second Edition, Springer, NY, 2004.

Robonaut, available at http://robonaut.jsc.nasa.gov/R2/.

Robonaut Performs Taskboard Tethering, available at http://www.nasa.gov/content/robonaut-video-gallery.

J. M. Robson, N by N checkers is Exptime complete, *SIAM Journal on Computing*, **13** (2) pp. 252-267.

J. M. Robson, The complexity of Go, *Information Processing: Proceedings of IFIP Congress*, pp. 413417.

K. Rosen, *Discrete Mathematics and its Applications*, Sixth Edition, McGraw Hill, NY, 2007.

P. Ross, "Why isn't there a Nobel Prize in Mathematics?" Available at http://mathforum.org/social/articles/ross.html.

B. Rozenblit, *Us Against Them: How Tribalism affects the Way We Think*, Transcendent Publications, MO, 2008.

S. Russell, D. Dewey, M. Tegmark, Research Priorities for Robust and Beneficial Artificial Intelligence, available at http://arxiv.org/abs/1602.03506.

C. Sagan, *Cosmos*, Random House, NY, 1980.

—, *Broca's Brain: Reflections on the Romance of Science*, Random House, NY, 1979.

A. Samuel, Some Studies in Machine Learning using the Game of Checkers, *IBM Journal of Research and Development*, **3** (1959), pp. 210229. doi:10.1147/rd.33.0210.

Arthur Samuel: Pioneer in Machine Learning, available at http://infolab.stanford.edu/pub/voy/museum/samuel.html.

Samuel's Checkers Player, available at http://webdocs.cs.ualberta.ca/˜sutton/book/ebook/node109.html.

N. Savage, When Computers stand in the Schoolhouse Door, *ACM Communications*, **59** 3 (2016) pp. 19-21.

W. R. Scott, *Group Theory*, Dover Publications, NY, 1987.

J. R. Searle, Minds, Brains and Programs, *Behavior and Brain Sciences*, **3**, 3 (1980) pp. 417–458.

Oliver Selfridge (Obit.), available at http://www.telegraph.co.uk/news/obituaries/3903053/Oliver-Selfridge.html.

O. G. Selfridge, Pattern Recognition in Modern Computers, *Proceedings of the Western Joint Computer Conference* (1955).

Mr. Selfridge, *Masterpiece* Drama, PBS, available at http://www.pbs.org/wgbh/masterpiece/programs/mr-selfridge/.

C. E. Shannon, A Symbolic Analysis of Relays amd Switching Circuits, Masters thesis, available at http://dspace.mit.edu/bitstream/handle/1721.1/11173/34541425-MIT.pdf.

Claude Shannon, Wiki article available at http://en.wikipedia.org/wiki/Claude_Shannon

M. W. Shelley, *Frankenstein, or the Modern Prometheus*, available at http://www.gutenberg.org/ebooks/84.

W. L. Shirer, *The Rise and Fall of the Third Reich: A History of Nazi Germany*, Simon and Shuster, NY, 1960.

A. Silberschatz, P. B. Galvin, G. Gagne, *Operating System Concepts*, Ninth Edition, John Wiley and Sons, Inc., MA, 2013.

Silent Running (Science fiction film, 1972), Wiki article available at http://en.wikipedia.org/wiki/Silent_Running.

Herbert A. Simon (Bio.), available at http://history.computer.org/pioneers/pdfs/S/Simon.pdf.

S. Singh, *Fermat's Enigma: The Epic Quest to solve the World's greatest Mathematical Problem*, Walker and Co., NY, 1997.

M. Sipser, *Introduction to the Theory of Computation*, PWS Publishing Co., Boston, MA, 1997.

Carl Sagan's Universe, edited by Y. Terzian, E. Bilson, Cambridge University Press, UK, 1997.

A. Smith, Hidden Markov Models and Mouse Ultrasonic Vocalizations, *XRDS*, **21** 4 (2015) pp. 48–54.

W. Stallings, L. Brown, *Computer Network Security: Principles and Practice*, Pearson Education, NJ,2008.

Theseus Maze Solving Mouse (invented by Claude E. Shannon). Material can be found on Shannon's invention (after a little navigating), at http://cyberneticzoo.com/mazesolvers/.

S. Thrun, *Probabilistic Robotics*, MIT Press, MA, 2005.

A. Turing, Computing Machinery and Intelligence, Mind, LIX (236): 433460, doi:10.1093/ mind/LIX.236.433. His seminal paper on the "Turing Test." Available at http:// m.mind.oxfordjournals.org/content/LIX/236/433.full.pdf.

—, On Computable Numbers, with an Application to the *Entscheidungsproblem*, *Proceedings of the London Mathematical Society*, **2** 42 (1936), pp. 230–265.

J. Wang, *Computer Network Security: Theory and Practice*, Springer, NY, 2009.

J. Weizenbaum, Eliza: A Computer Program for the Study of Natural Language Communication between Man and Machine, *ACM Communications*, **9** 1 (1966), pp. 36–45, available at http://web.stanford.edu/class/linguist238/p36-weizenbaum.pdf.

Norbert Wiener, Wiki article available at http://en.wikipedia.org/wiki/ Norbert_Wiener

N. Wiener, *Cybernetics: Or Control and Communication in the Animal and the Machine*, Hermann and Cie (Paris), MIT Press (MA), 1948.

W. G. Walter, Wiki article available at http://en.wikipedia.org/wiki/ William_Grey_Walter.

—, *An Electromechanical Animal*, *Dialectica*, **4** (1950) pp. 42–49.

D. A. Wilson and E. O. Wilson, Evolution "for the good of the Group," *American Scientist*, **96** 5 (2008) pp. 380–389.

E. O. Wilson,*On Human Nature*, Harvard University Press, MA, 2004.

W. A. Wilson, J. I. Tracey, *Analytic Geometry*, DC Heath and Co., NY, 1937.

Why is Lisp used for AI? Available at http://stackoverflow.com/questions/ 130475/whyislispusedforai.

D. Zarrouk, M. Mann, N. Degani, T. Yehuda, N. Jarbi, A. Hess. Single actuator wavelike robot (SAW): design, modeling, and experiments. *Bioinspiration and Biomimetics*, 11 (4): 046004 DOI: 10.1088/1748-3190/11/4/046004.

Index